Solar Wind

Solar Wind

Edited by **Catherine Waltz**

New York

Published by Callisto Reference,
106 Park Avenue, Suite 200,
New York, NY 10016, USA
www.callistoreference.com

Solar Wind
Edited by Catherine Waltz

International Standard Book Number: 978-1-63239-572-6 (Hardback)

Printed in the United States of America.

Contents

Preface

This book has been a concerted effort by a group of academicians, researchers and scientists, who have contributed their research works for the realization of the book. This book has materialized in the wake of emerging advancements and innovations in this field. Therefore, the need of the hour was to compile all the required researches and disseminate the knowledge to a broad spectrum of people comprising of students, researchers and specialists of the field.

This book contains the work of prominent scientists in the fields of solar and space plasma physics, and space and planetary physics. It makes a significant contribution to the theory, modeling and experimental methods of the solar wind exploration. The main objective of the book is to provide the latest knowledge regarding solar wind formation and elemental composition, the interplanetary dynamical evolution and acceleration of the charged plasma particles, and the guiding magnetic field that connects to the magnetospheric field lines and that adjusts the effects of the solar wind on the Earth. Many of the scientists actively working and researching in these fields will find this book full of several new and interesting ideas.

At the end of the preface, I would like to thank the authors for their brilliant chapters and the publisher for guiding us all-through the making of the book till its final stage. Also, I would like to thank my family for providing the support and encouragement throughout my academic career and research projects.

Editor

Part 1

The Solar Wind Magnetic Field
Powered by the Sun

Impact of the Large-Scale Solar Magnetic Field on the Solar Corona and Solar Wind

A.G. Tlatov[1] and B.P. Filippov[2]
*[1]Kislovodsk Mountain Station of the Central
Astronomical Observatory of RAS at Pulkovo
[2]Pushkov Institute of Terrestrial Magnetism,
Ionosphere and Radio Wave Propagation,
Russian Academy of Sciences, Troitsk, Moscow Region
Russia*

1. Introduction

In 1955 Soviet astrophysicists Vsehsvyatskiy, Nikolskiy, Ponomarev and Cherednichenko (Vsekhsvyatskiy et al., 1955) showed that broad corona loses its energy for radiance, and can be in hydrodynamic equilibrium, and there should be a flow of materials and energy.

This process is a physical basis for the important phenomenon of "dynamic corona". The magnitude of the flow of materials was evaluated due to the following considerations: if the corona were in hydrodynamic equilibrium, then the altitudes of homogenous atmosphere for hydrogen and iron would correlate as 56/1. In other words, in such case iron ions must not be observed in the distant corona. But this is not so. In 1955 it was a considerable achievement, but nobody believed in the phenomenon of "dynamic corona".

Three years later Eugene N. Parker came to the conclusion that hot solar stream in Chapman model and particle flux, blowing away commentary tails in Birmann's hypothesis – these are manifestation of the same phenomenon, and Eugene N. Parker called it "solar wind".

Parker (Parker, 1958) showed - despite the fact that solar corona is greatly gravitated to the Sun, it is a strong heat conductor, it remains hot even at great distance. The farther the distance, the less the solar gravitation is, there is a supersonic discharge from the upper corona into interplanetary space.

Solar wind represents a flux of ionized particles, thrown out of the Sun in all the directions with the speed about 300-1200 km/sec. The source of the solar wind is solar corona. The temperature of the solar corona is so high, that gravitation force is not able to hold its substance near the surface, and part of this substance constantly moves to interplanetary space.

First direct gaging of the solar wind was carried out in 1959 by the automatic interplanetary station "Luna-1". The observations were made by means of a scintillometer and a gas ionization detector. Three years later the same gaging was implemented by the American scientists on board the station "Mariner-2".

First numerical models of solar wind in corona with using the equations of magnetofluid dynamics were created by Pneuman and Kopp in 1971 (Pneuman & Kopp, 1971).

In the end of the 90s there were observations of the areas of uprise of fast solar wind in solar poles, those observations were made on board of the satellite SOHO by means of Ultraviolet Coronal Spectrometer (UVCS). It turned out that acceleration of the wind is much higher than it was presupposed, judging from purely thermodynamic extension. Parker's model predicted that the wind speed becomes hypersonic in 4 solar radii altitude from photosphere, but the observations showed that this transition takes place significantly lower, approximately in 1 solar radius altitude, confirming the existence of the extra mechanism of acceleration of the solar wind.

Long-term observations from the orbit of the Earth (about 150 000 000 km distance from the Sun) showed that the solar wind is structured and can usually be classified as steady and perturbative (sporadic and recurrent).

Depending on the speed, sporadic streams of solar wind can be divided into slow (approximately $300-500$ km/sec near the orbit of the Earth) and fast ($500-800$ km/sec near the orbit of the Earth) (Fig. 1).

Slow solar wind is generated by the "steady" part of the solar corona (area of coronal streamers) with its gas-dynamic extension: with the temperature of corona about 2×10^6 K – cjrona cannot be in the condition of hydrostatic equilibrium, and this extension under the present border conditions must lead to the acceleration of the corona substance up to hypersonic speeds.

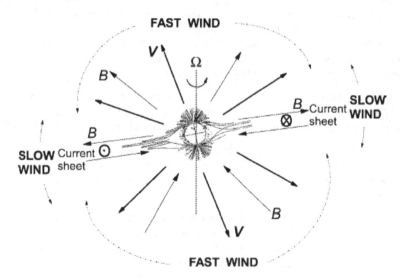

Fig. 1. Simplified picture of the large-scale structure of the solar wind near sunspot minimum, when the solar magnetic dipole makes a small angle with the spin axis (dotted line). The velocity and field lines are sketched in bold and thin lines respectively. The magnetic polarity is the one that existed when the WIND and Ulysses observations were acquired; this polarity reverses every 11 years (Meyer-Vernet, 2007).

Solar corona heating up to such temperatures takes place due to convectional nature of heat transfer in the solar photosphere: the development of the convective turbulence in plasma is accompanied by the generation of intensive magnetoacoustic waves. In its turn, during transmission towards solar atmosphere density reduction, sound waves are transforming into shock waves; shock waves are effectively absorbed by the corona substance and heat it till the temperature of $(1-3)\times10^6$ K. The streams of recurrent fast solar wind are emanated by the Sun during several months and have a return period of 27 days (the period of the rotation of the Sun) when being observed from the Earth. These streams are associated with coronal holes – the areas of the corona with relatively low temperature (approximately $0,8\times10^6$ K), low-density of plasma (that is all in all a quarter of density of the sporadic corona areas) and radial in relation to the Sun magnetic field. Owing to high conductivity of the solar wind plasma magnetic field of the Sun turns out to be frozen-in into the flowing out wind streams, and can be observed within the interplanetary atmosphere as an interplanetary magnetic field.

Because of the solar wind the Sun loses about one million tons of its substance every second. The solar wind consists basically of electrons, protons hellions (alpha-particles); nuclei of other elements and non-ionized particles (electrically neutral) are contained in rather inconspicuous quantity.

2. Long-term changes of coronal shape and geomagnetic disturbance

Solar wind parameters are changing during the solar activity. Their direct determination has begun relatively recently, with the beginning of space age. But the observations of the geomagnetic disturbance and coronal shape give us the opportunity to evaluate variations of the solar wind throughout the period of more than 100 years.

The large-scale solar corona structure corresponds to the large-scale configuration of solar magnetic fields. Since the magnetic field of the Sun is subjected to cyclic variations, the coronal shape also changes cyclically. Processing 12 photographs of the corona during solar eclipses, Ganskiy (Ganskiy, *1897*) classified 3 types of corona, i.e., maximum, intermediate, and minimum. In 1902, in the report concerning the solar eclipse of 1898, Naegamvala (*Naegamvala, 1902*) also gave a corona classification that depended on the sunspot activity. Description of the corona shape involves the use of characteristic features and the phase of solar activity that is given by $\Phi = \dfrac{T - T_{min}}{|T_{max} - T_{min}|}$. The values of Φ are positive and negative at the rising and declining branches of the solar cycle. Vsekhsvyatskiy (Vsekhsvyatskiy et al., 1965) gave a somewhat different classification of the structure types. They are (1) a maximum type $\Phi > 0.85$ in which polar-ray structures are not seen, large streamers are observed at all heliolatitudes and are situated radially; (2) an intermediate premaximum or postmaximum type $0.85 > |\Phi| > 0.5$, in which polar-ray structures are observed at least in one hemisphere, and large coronal streamers situated almost radially are clearly seen at high latitudes; (3) an intermediate preminimum or postminimum type $0.5 > |\Phi| > 0.15$ in which polar-ray structures are clearly seen in both hemispheres and large coronal streamers strongly deviate toward the solar equator plane; (4) a minimum type $0.15 > |\Phi|$ in which polar-ray structures are clearly seen in both hemispheres and large coronal streamers are parallel to the equator plane; and (5) an ideally minimum type $0.05 > |\Phi|$ in which powerful structures of large coronal streamers are situated along the equator. Changes in the extent of the polar-ray structures, the degree of corona flattening, the average angle between large coronal streamers, and other characteristics of the corona that depend on the solar cycle phase have been widely studied (Loucif and Koutchmy, 1989, Vsekhsvyatskiy et al., 1965, Golub & Pasachoff, 2009).

Solar eclipses of 2006 Mar 26, 2008 Jul 22, and 2009 Aug 1 have enabled detailed examination of changes in the shape of the corona in the period of minimum solar activity in the modern era. However, the shape of the corona in the current minimum is slightly different from the ideal shape of the eclipses in the minimum of activity.

This section contains the comparative analysis of the solar corona structure during the minimums of activity cycles 12-24, and their changes throughout the centuries are discussed.

2.1 Processing method and results

The initial data in the analysis were drawings of the corona shape taken from the catalogs of (Vsekhsvyatskiy et al., 1965) (see Fig. 3, 1-8), Loucif and Koutchmy, 1989) (Fig. 3, 9) and also drawings of the eclipses at minima of cycles 21 and 23 taken from Waldmeier (Waldmeier, 1976) (Fig. 3, 10), Gulyaev (Gulyaev, 1997) (Fig. 2, 11) (Eclipse photo of 01.08.2008 with permission of M. Druckmüller), (Fig. 2, 12). (Vsekhsvyatskiy et al., 1965) separated out an ideally minimum corona. It is supposed that the cycle phase must be $|\Phi| < 0.05$. In fact, the ideally minimum corona type was observed on in 1954 (see Fig. 3). This happened at the solar activity minimum before the largest one in the history of observation cycle 19. Most probably, such a corona type did not occur for 50 years before and after this event. The corona shape of 1954 was close to a dipole one. This means that large coronal streamers rapidly approach the solar equator plane. However, during other eclipses, such as 1889 Jan 1 ($\Phi = -0.18$), 1889 Dec 21 ($\Phi = 0.03$), 1901 May 17 ($\Phi = -0.07$), 1923 Sep 10 ($\Phi = 0.04$), and 1965 May 30 ($\Phi = 0.14$), the large coronal streamers at distances more than 2 solar radii do not lie on the equator plane. They expand from mid-latitudes in parallel or at a slight angle to the equatorial plane (see Fig. 3). To analyze the corona shape at the eclipses during a solar activity minimum epoch, a corresponding index should be chosen. We need the index that characterizes the shape of the corona of the minimal type and is applicable to images and drawings of different qualities. The corona of the solar minimum is characterized by pronounced polar ray structures and large coronal streamers (Pasachoff et al., 2008). Let us introduce the parameter γ that characterizes the angle between high-latitude boundaries of the large coronal streamers at a distance of $2 \cdot R_\Theta$. The γ parameter is the sum of the angles at the eastern and western limbs: $\gamma=180-(\gamma_W +\gamma_E)$. Figure 2 gives the scheme showing how parameter γ is determined from the shapes of the corona in Fig. 3, and the error does not exceed 4⁰.

The parameter γ is close to the index of the extent of the polar regions (Nikolskiy 1955; Loucif & Koutchmy, 1989; Golub, 2009). But in comparison with the polar index (Loucif & Koutchmy, 1989), it is not calculated close to the limb, but at a height R_Θ above the solar limb. This allows taking the compression of coronal rays to the plane of which is the equator into account, possibly linked with non-radial spreading of the coronal streamers. This is especially significant for the minimum activity epoch (Tlatov, 2010a).

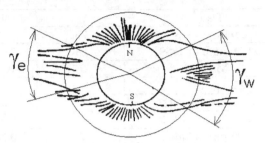

Fig. 2. Scheme showing how the angles defining parameter $\gamma = 180 - (\gamma_W + \gamma_E)$ are found for the eclipse of 1923. The outer thin circle has a radius of $2R_o$

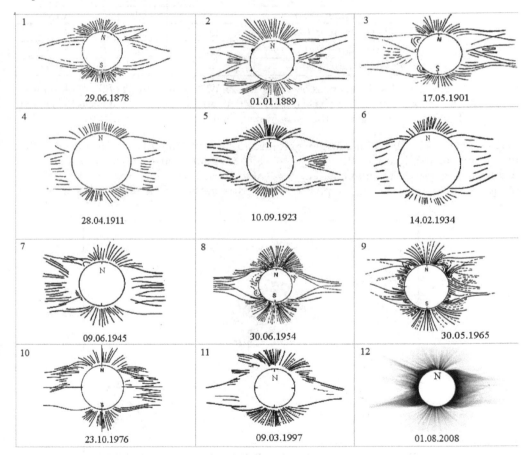

Fig. 3. Eclipses close to the minima of cycles 12--24.

Figure 3 presents the shapes of eclipses for the epochs close to the solar activity minima of cycles 12-24. The calculated values of the parameter γ for these eclipses are listed in Table 1. The γ parameter varies within $40 \div 100$ degrees. Table 1 also gives the solar cycle phase Φ (see Eq. 1). One can see that the highest magnitudes of the parameter γ occurred during the

Cycle No	Date	W	Φ	γ
12	29 June 1878	75	-0.06	65
13	21 December 1889	88	0.03	60
14	17 May 1901	63	-0.07	80
15	28 April 1911	103	-0.18	80
16	10 September 1923	77	0.04	85
17	14 February 1934	114	0.14	97
18	09 June 1945	151	0.28	88*
19	30 June 1954	190	0	98
20	30 May 1965	106	0.14	90
21	23 October 1976	155	0,08	67
22	22 November 1984	158	-0.35	69*
23	09 March 1997	125	0.1	60
24	01 August 2008	--	~0.1	54

Table 1. Parameters γ for the eclipses of cycles 12 - 24. W - the amplitude of the Wolf number at the sunspot maximum, Φ- the phase of solar activity for eclipse date. For cycle 18 and 22 γ was modified taking the phase of the cycle into account

period 1934--1955. The remaining magnitudes fit the enveloping curve fairly well with a maximum during cycles 17--19 (see Fig. 4). No information on the solar corona structure during the eclipses at the minima of cycles 18 and 22 has been found in the literature. To fill the gaps, the eclipses during the phases of growth or decline of the solar cycle can be used.

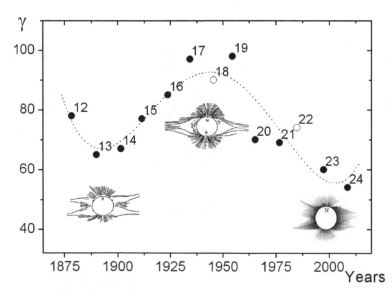

Fig. 4. Distribution of parameter γ for the structure of the corona of the minimal type. Numbers of activity cycles are given. The shape of the solar corona to cycles 13,19 and 24 are present.

One can see in Table 1 that the eclipses of 1945 and 1984 are rather far from the minimum phase in solar activity. A modified parameter $\gamma^*=180-(\gamma_W +\gamma_E) \cdot (1-| \Phi |)$ can be introduced for these eclipses. This parameter reduces parameter γ to the minimum phase. The application of this procedure is effective for recognizing the shape of the corona close to the minimum activity with the phase $\Phi <0.4$. Figure 4 presents variations in parameter γ during the last 13 activity cycles.

2.2 Long-term changes of coronal shape and geomagnetic disturbance

The presence of long-term trends in the solar corona structure can be caused by changes in the configuration of the global magnetic field of the Sun. The role of active region formation during a solar activity minimum is not significant. It has long been known that large coronal streamers typically lie above the polarity-inversion lines of the large-scale magnetic field marked by filaments and prominences (Vsekhsvyatskiy et al., 1965). For this reason, investigations of the large-scale corona shape give valuable information on the structure of the large-scale fields during a long time interval. During the activity minimum, the properties of the global magnetic field of the Sun manifest themselves in the most pronounced way.

The magnetic field of the Sun is determined by large-scale structures. The northern and southern hemispheres of the Sun have magnetic fields of opposite polarity. The strength of the polar magnetic field is significantly higher than the fields in middle and low latitudes in the activity minimum period. Along with this, one can conclude from the analysis that assuming that the global solar field configuration is in the form of a dipole structure is probably incorrect. The corona configurations for the eclipses of 1889, 1901, and others correspond instead to the quadrupole form, or to an octopole form if different polarities at the poles are taken into account.

Thus, long-term variations in parameter γ should manifest changes in the dipole component during solar activity minima. This hypothesis can be checked using the data on configurations of the large-scale magnetic fields. Figure 5 shows changes in the dipole moment and the envelope drawn through solar activity minima. The data were obtained from the analysis of synoptic H-alpha charts of patterns of polarity inversion lines from (Makarov & Sivaraman, 1989; Vasil'eva, 2002), and the Kislovodsk Astronomical Mountain Station (http://www.solarstation.ru). The greatest dipole moment corresponded to the minimum of cycle 19 in 1954. These data were obtained from *H-alpha* synoptic maps. The amplitude of the dipole contributions $l=1$, which determine parameter A_1 depend on the topology of the large-scale magnetic fields. The method of decomposition is given in the articles (Makarov & Tlatov, 2000; Tlatov, 2009). On the whole, the envelope of the dipole moment shows similar trends in the changes in the corona shape parameter γ (see Figs. 4 and 5).

The growth in the strength of the radial component of the interplanetary magnetic field (Cliver, 2002) determined from the geomagnetic activity index *aa* is another important problem that has been widely discussed recently. Figure 6 shows variations in the geomagnetic index *aa*. Data were taken from the National Geophysical Data Center (http://www.ngdc.noaa.gov). The first half of the 20th century was characterized by a growth in this index; i.e., the slowly varying component that was especially pronounced

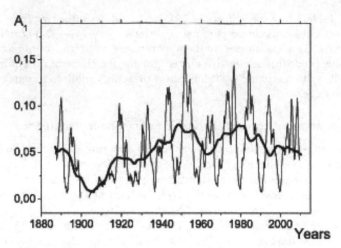

Fig. 5. Variations in the dipole moment derived from synoptic H-alpha charts of the Sun
$$A(t) = \sum_{m,l=1} (g_l^m g_l^m + h_l^m h_l^m), \, l=1, m=0,1,$$ where g_l^m and h_l^m are the coefficients of the spherical function expansion. The thin line indicates monthly values, and the thick line are 11-year running means.

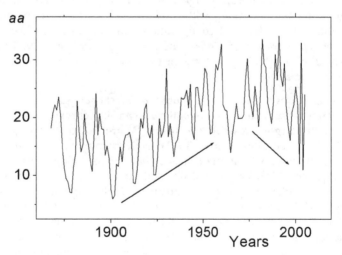

Fig. 6. Annual mean of the geomagnetic indices *aa* from 1868 (according to the National Geophysical Data Center, NGDC), smoothed by 2 years. The arrows mark the index growth in the first half of the 20th century and the index decrease during the last decades for the solar activity minimum epochs.

during the activity minimum epochs grew. During the past decades, a decrease in the *aa* index during the activity minimum epochs was observed. Probably, this is due to rearrangement of the global magnetic field of the Sun accompanied by changes in the solar corona structure during the minimum epochs. Variations in the geomagnetic index and the dipole moment of the large-scale magnetic field of the Sun during the activity minimum epoch are almost identical (see Fig. 7).

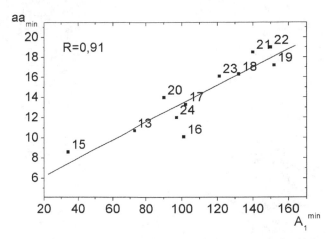

Fig. 7. Relation between the geomagnetic index aa and the magnitude of the dipole moment of the large-scale magnetic field A_1 according to Figs. 4 and 5 during the solar activity minimum epoch. Numbers of activity cycles and linear regression are also shown.

We can also test changes deviations of coronal streamers from the radial direction for over 100 years. You can use the catalogs of solar eclipses (Loucif and Koutchmy, 1989; Naegamvala, 1902; Vsekhsvyatskiy et al., 1965).

The corona of the solar minimum is characterized by pronounced polar ray structures and large coronal streamers. Let us introduce index γ that characterizes the angle between high-latitude boundaries of the large coronal streamers at a distance of 2R. The γ index is the sum of the angles at the eastern and western limbs: $\gamma = \gamma_N + \gamma_S$. In fact, the γ index is a simpler version of the corona flattening indices (Nikolskiy, 1955) but it is calculated at height R above solar limb. Figure 4 presents variations in parameter γ during the last 13 activity cycles (Tlatov, 2010a).

Thus, analysis of the corona shape has revealed a long-term modulation of the global magnetic field of the Sun. Possibly, there exists a secular modulation of the global solar magnetic field which is most pronounced during the solar activity minimum epoch. During the secular cycle of the global magnetic field of the Sun the relation between the dipole and quadrupole components of the magnetic field changes. The largest amplitude of the dipole component occurred during the interval 1944--1955. At the boundary between the 20th and 21st centuries the solar corona shape and, possibly, the global magnetic field correspond to the configuration close to the octupole one (Fig. 8).

3. Non-radial spreading of the solar corona structures in solar cycle and variations of the solar wind parameters

In this section we will consider in detail the physical mechanism linked with variations of the coronal shape and changes of the solar wind parameters.

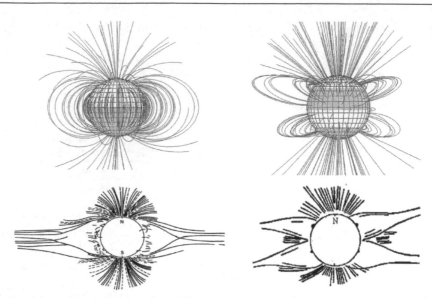

Fig. 8. Configuration of magnetic force lines for the corona of the dipole type (left) and quadrupole or octopole type if different polarities at the poles are considered (right). Regions of negative polarity are darkened. The structure of the corona for the epoch of minimum 13 and 19 activity cycles are also shown.

The solar corona structure corresponds to the configuration of solar magnetic fields. Since the magnetic field of the Sun is subjected to cyclic variations, the corona shape also changes cyclically. The coronal rays are distinctive structures in the solar corona, which propagate at a small angle to the radial direction from the Sun and display the electron density in K-corona enhanced by the factor of 3 to 10. The angle $\Delta\theta$ that describes the deviation of the rays from the radial position varies with the phase of the solar cycle and the latitude (Eselevich and Eselevich, 2002; Tlatov 2010b).

The regular observations with the SOHO/LASCO and Mark-III/IV coronagraphs at the Mauna Loa solar observatory make it possible to analyze the structure of the solar corona for the time comparable with the duration of the solar cycle. These data substantially complement extended series of observations of the corona in spectral lines carried out with extra-eclipse coronagraphs, since they make it possible to analyze coronal structures at sufficiently large distances from the solar limb, and also occasional observations of the "white light" corona during total eclipses. The coronagraph -- polarimeter Mark III detected the structure of the solar corona at the heights ~ 1,15 ÷ 2,45 $R_\odot$ in 1980-1999. In 1998, at Mauna Loa observatory the new low-noise coronagraph Mark-IV, with a liquid-crystal modulator of polarization and a CCD, was mounted. To decrease the radial gradient and consequently to increase the contrast, we applied to the Mark data an artificial vignetting function. The LASCO-2 coronagraphic telescope on board of SOHO satellite has been working since 1996 and covers the distance 1.5 ÷ 6 $R_\odot$ above the solar limb. Thereby, here we have analyzed the structure of the corona for 1980-2008 on the basis of the data obtained at ground-based observations with Mark-III/IV coronagraphs and for 1996-2009 with the SOHO/LASCO-2 data.

3.1 The identification for the deviation angles of coronal rays

In order to determine the deviation of coronal rays, we developed a technique of the identification of coronal streamers in two-dimensional images of the corona obtained with SOHO/LASCO-2 and Mark-III/IV in automatic mode. The analysis is based on discrimination of central parts of bright coronal structures propagating, as a rule, at some angle to the radial direction, discrimination of the points of the local maxima, and determination of the parameters of the approximating line section (Figure 9). The procedure included the following stages. Initially, the coordinates of the center and the radius of the Sun were measured in pixels. Further on, we calculated the average limb brightness of the corona for different heights above the solar limb $I(r)$. We considered the regions in which the brightness at a given distance from the limb was no smaller than $0.3 \cdot I(r)$. Then we selected all the points with the brightness exceeding that of the points apart of it by the angle $\pm 1.5°$ along the limb. The obtained collections of points, as a rule, represented regions close to the center of brightness of the coronal rays, extended from the limb of the Sun at some angle to the radial direction. For these regions, we inscribed the least-square approximating straight lines. From the parameters of these straight lines we determined the polar angle of the base of a coronal ray θ and the deviation from the radial direction $\Delta\theta$. Figure 10 presents an example of such identification for a Soho/Lasco-2 image obtained on 2007.03.16. We used for the analysis the deviation angle for these straight lines with respect to the radial propagation. We took into account line sections with the length no smaller than $1.0 \cdot R$ for SOHO/Lasco-2 images and than $0.5R$ for Mark-III/IV images. The selection of rays, the linearity of which can be traced to a fairly large distance, allows to filter out a considerable number of coronal structures associated with eruptive processes. Thereby, we have processed about $4.2 \cdot 10^3$ images for the time interval 1996-2009 and discriminated approximately 10^5 coronal rays from the Soho/Lasco-2 data and $7 \cdot 10^3$ days for the Mark-III/IV data for 1980-2008.

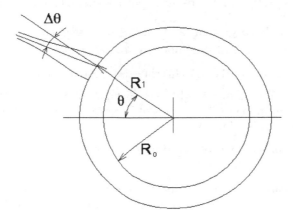

Fig. 9. The determination of deviation angles for coronal rays.

In this analysis, we recorded the coronal beams of various types. Among the rays belonging to the helmet-type, and chains of streamers, rays belonging to the low-latitude and polar regions (*Saito et al. 2000, Eselevich & Eselevich, 2002*). These rays can be formed over the bipolar and unipolar magnetic structures. There are also dynamic coronal rays, but we

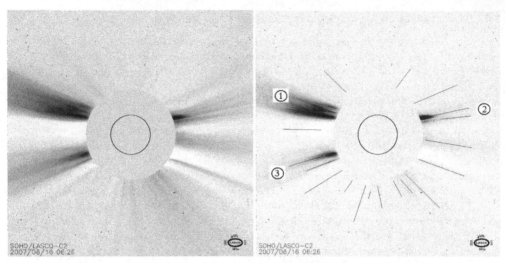

Fig. 10. An example of the identification of the direction of the propagation of coronal streamers for an image of the corona obtained with Soho/Lasco-2 on 2007.03.16.

decided that their number is small enough to that would affect the comparison. Thus Fig. 11 shows that the footpoints rays are usually evident in a few days. In addition to the geometry of the rays in the process of selection, we also recorded the absolute and relative brightness of the rays. This allowed to make a comparative analysis nonradially rays and define their types. In order to define the brightness of rays we calculated the average intensity within the function from the height over the solar limb

$$I_{avr}(r) = 1/2\pi \int_{0}^{2\pi} I_{avr}(r,\alpha)d\alpha \,,$$ where α- is a polar angle. Then we defined the average

intensity of a ray at the beginning and ending of a segment, approximating the ray $\overline{I} = (I_{beg} + I_{end})/2$. The rays with the intensity $\overline{I} > 2 \cdot I_{avr}$, where I_{avr} was calculated at a

height $(R_{beg} + R_{end})/2$, were related to bright rays. The rays with the intensity $\overline{I} < 1.3 \cdot I_{avr}$ were conditionally related to non-bright rays. Figure 10 shows the systems of bright rays, that are marked with numbers 1-3. The rays of lower intensity are normally located within the area of high latitudes (4, Fig. 10).

Footpoints of bright coronal rays, obtained under the assumption of linear distribution, as a rule, close to the neutral line (Fig. 11).

Fig. 12 shows the change of nonradiality parameter for bright rays $\Delta\theta_{br}$ associated with helmet-rays and soft rays $\Delta\theta_{low}$, which are typical for the chain of coronal streamers from the unipolar regions. Rays of varying brightness show close cyclical course of the parameter $\Delta\theta$, although the degree of nonradiality for bright rays is slightly less: $\Delta\theta_{br}=1.46+0.47\cdot\Delta\theta_{low}$, are correlated with $r=0.79$. Rays of different latitudinal zones also show a close behavior (Fig. 13). Relationship between the parameter $\Delta\theta$, for rays of the equatorial and polar zones is the following: $\Delta\theta_{pol}=4.1+0.84\cdot\Delta\theta_{eq}$; $r=0.89$. Thus, the nonradial parameter $\Delta\theta$ cycle exists for the various types of coronal rays.

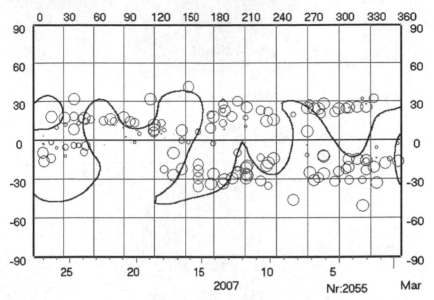

Fig. 11. Example of synoptic maps for Carrington rotation N2055 which marked the neutral line, calculated at an altitude $R = 1.9 \cdot R_\odot$ according to the Wilcox Solar Observatory. Presents the footpoints of bright coronal rays.

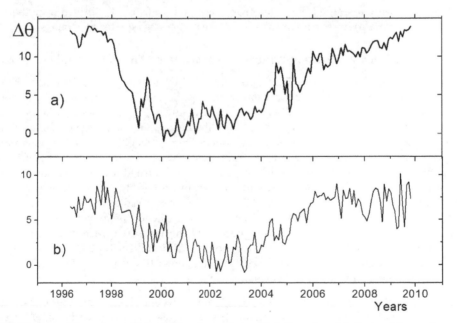

Fig. 12. The monthly-averaged deviation from the radial direction $\Delta\theta$ derived from the SOHO/Lasco-2 data a) for coronal rays of low intensity $I<1.3 \cdot I_{avr}$. b) for bright coronal rays ($I>2 \cdot I_{avr}$).

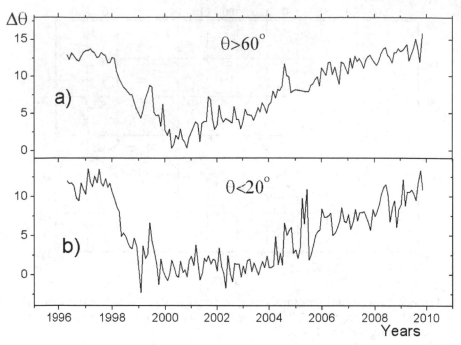

Fig. 13. The monthly-averaged deviation from the radial direction $\Delta\theta$ derived from the SOHO/Lasco-2 data a) for the polar rays b) for the rays of the equatorial zone.

3.2 Connection between the non-radial coronal shape and the solar wind parameters

The analysis of the SOHO/Lasco-2 data for 1996-2009 indicates that the deviation angle $\Delta\theta$ varies with the 11-year cycle of the activity, reaching the maximum values at the minimum of the activity. Figure 14 presents the time-latitude diagram for $\Delta\theta$ variations. In the vicinity of the equator, the deviation of the rays from the radial direction reaches the minimum. Negative angles were seen in the northern hemisphere in 2002-2004 and in the southern hemisphere in 2000-2003, at the latitudes smaller than 30°. The coronal rays at the minimum of the activity and the phases of the decline are, as a rule, turned towards the solar equator. Only rather low-latitude coronal rays at the maximum of the activity slightly deviate towards the poles (Fig. 14). The largest deviation from the radial direction is seen at the minimum of the activity at the latitudes 30 ÷ 60°. At the time of the maximum of the activity and the polarity reversal of the solar magnetic field the rays are directed either parallel to the equator or slightly deviate towards the poles. Individual fluctuations of the angles of deviation of the rays are seen in different latitude zones (Figure 13), which indicates the general type of the perturbations of coronal structures.

Figure 15 presents the monthly averages for the deviation, averaged along the entire limb for all types of coronal rays. For comparison, the graph also shows the variation of the angle of tilt angle τ of the heliospheric current sheet (HCS) (Hoeksema & Scherrer, 1986). Tilt angle HCS is calculated in a potential assumption of this photospheric magnetic field for $R=3.25 \cdot R_0$. Between these parameters there is the ratio: $\Delta\theta = 0.49 + 0.2 \cdot \tau$; $r = 0.91$.

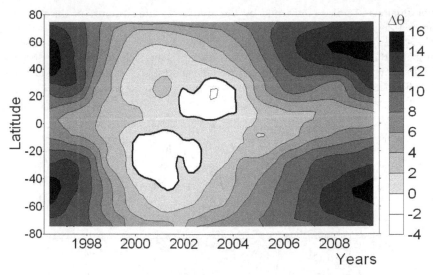

Fig. 14. The latitude and time distribution of the angles of the deviation of coronal rays from the radial direction in the time interval 1996-2009 derived from the SOHO/Lasco-2 data. The regions of the deviation of the rays towards the equator are darkened. The levels are indicated at the intervals of 2 degrees; the level corresponding to $\Delta\theta = 0°$ is shown by the thick line.

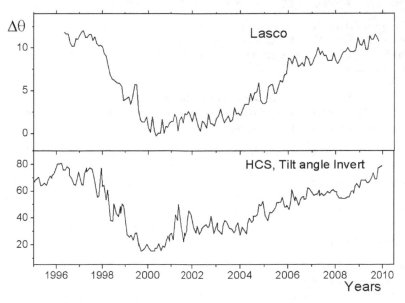

Fig. 15. Comparison between the deviation $\Delta\theta$ of coronal rays from SOHO/Lasco-2 (top panel) and the inverted tilt angle of the heliospheric current sheet according to WSO (http://wso.stanford.edu).

These conclusions are also confirmed by the analysis of the structure of the solar corona at the heights 1.2 ÷ 2.5R based on the MLSO Mark-III/IV data. Figure 16 presents the angle $\Delta\theta$ for the latitude zone $\pm 30^{\circ}$ in the time interval 1980-2008. For the period 1996-2008 angle $\Delta\theta$ according to the coronagraph Mark was slightly lower than according to Lasco/C2: $\Delta\theta_{Mark}=2.55+0.66 \cdot \Delta\theta_{Lasco}$; $r=0.95$. The nonradial variation $\Delta\theta$ according to Mark telescope during 1980-2008 change of the parameter also has a good correlation with the angle of tilt angle of HCS $r\sim 0.8$ (Fig. 16).

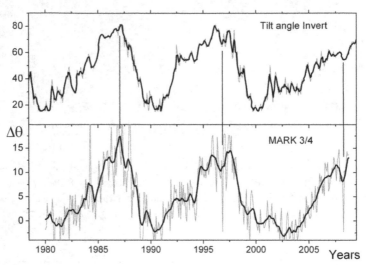

Fig. 16. The comparison between the deviations $\Delta\theta$ of coronal rays located in the middle-latitude zone $\pm 30^{\circ}$, from the MLSO- MarkIII/IV data (bottom panel) and the inverted tilt angle of the heliospheric current sheet according to the WSO (top panel). The values are smoothed for 6 months.

Figure 17 presents the latitude distribution for the $\Delta\theta$ angle for the time of the minimum of the 23-rd cycle. The maximum deviation is seen at the latitudes ~ 40-50°, which corresponds to the data obtained with the Lasco telescope (Eselevich & Eselevich, 2002).

The deviation of coronal rays from the radial direction $\Delta\theta$ related to the solar activity cycle may affect substantially the formation of the solar wind and geomagnetic perturbations. From geometrical consideration, in Figure 1 the flux of the solar wind at the distance r is related to that at the distance R_1, where the current lines become straight, as follows: $nv / n_1 v_1 = 1 + (\theta - \Delta\theta) \cdot (r - R_1) / \theta \cdot R_1$, for $\Delta\theta < \theta$. Assuming that at the years of the minimum activity $\theta = 40^{\circ}$ and $\Delta\theta = 20^{\circ}$ (Figure 9), we obtain that the ratio $nv / n_1 v_1$ at the distance 1 AU increases roughly twice compared to the case of radial expansion.

Figure 18 presents the graph of the deviation angle in comparison with the indexes connected with parameters of a solar wind. In the years of the maximum activity, the influence of active regions and flares is substantial. However during a minimum of activity the deviation $\Delta\theta$ to a plane of solar equator can be one of the reasons of delay of a solar wind speed (Figure 18c), and to modulate indexes of geomagnetic activity *aa* (Figure 18d) and cosmic rays intensity (CRI) (Figure 18b).

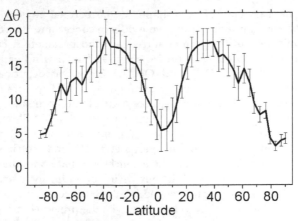

Fig. 17. The variations of the angle Δθ as a function of the latitude, obtained from the MLSO-MarkIII data at the minimum of the 23-rd cycle of the solar activity (1996-1997).

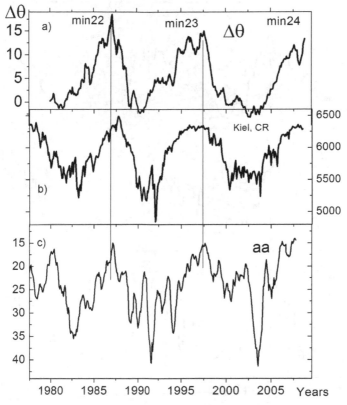

Fig. 18. The comparison between a) the deviations of coronal rays Δθ; b) data of Kiel cosmic-ray intensity (arbitrary units, monthly means); d) the geomagnetic index *aa*. The data are smoothed with a sliding window technique for 6 months.

The similarity of the behavior of the shape of the coronal rays with the indices of geomagnetic activity and the radial flux give evidence concerning the connection between the large-scale organization of the corona and solar wind parameters. But is it possible to establish the link directly between the coronal shape and solar wind parameters, measured from the Earth? Data bases OMNI1 and OMNI2 (http://nssdc.gfc.nasa.gov/omniweb) contain the information concerning hourly average value of key parameters of the solar activity, interplanetary atmosphere and geomagnetic disturbance since 1964. We collated changes of $\Delta\theta$ parameter with solar wind parameter $\beta = 8\pi nkT / B^2$ and magnetic Mach number $M_a = v / v_a = v\sqrt{\mu\mu_o\rho} / B$ (Fig. 19). In minimal activity epoch there is a correspondence between the non-radial parameter $\Delta\theta$ and solar wind parameters. Comparing parameters β и M_a we can make a conclusion that non-radial corona influences the relation ρ / B^2 of the solar wind, which grows in minimum solar activity and therefore magnetic field in minimal activity squeezes the solar wind flux towards helio-equator. These results complement the findings of in comparison with the previous three minima, this solar minimum has the slowest, least dense, and coolest solar wind, and the weakest magnetic field (Jian et al., 2011).

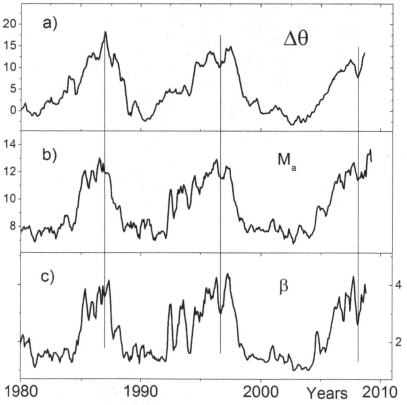

Fig. 19. The comparison between a) the deviations of coronal rays $\Delta\theta$ with the solar wind parameters according to OMNI2 database b) Alfven mach number Ma and c) Plasma beta β. Data of the solar wind parameters were are smoothed on 20 Bartels rotations.

4. Conclusion

Nowadays to prognose the solar wind parameters empirical models are used (Wang, Y.-M., & Sheeley, 1990; Arge & Pizzo, 2000) , they are based on the partial understanding of the physical processes taking place on the Sun. At the heart of these ideas there is a hypothesis about the radial character of the solar wind spreading. The details of the streams formation of the solar wind are complicated and are the subject of many researches. Nowadays there aren't any exact mathematical models of the physical processes influencing the formation of the solar wind flux.

In our research we pay attention to the fact that in minimal activity the conditions of forming the solar wind can depend on the configuration of the large-scale field and the shape of the solar corona. The configuration of the global magnetic field in minimal activity corresponds to the low harmonics of decomposition according to the Legendre polynomials $l=1,2,3$.

The link between the coronal shape, geomagnetic disturbances and solar wind parameters can be clearly seen in 11-year activity cycle. Global magnetic field can lead to non-radial spreading of the coronal streamers and probably, to the spreading of the solar wind. In minimal activity when the value of the global field reaches its maximum, coronal rays are deflected aside from the helio-equator.

Thus, analysis of the corona shape has revealed a long-term modulation of the global magnetic field of the Sun. Possibly, a secular modulation exists of the global solar magnetic field that is most pronounced during the solar activity minimum epoch. During the secular cycle of the global magnetic field of the Sun, the relation between the dipole and octopole components of the magnetic field changes. The largest amplitude of the dipole component occurred during the interval 1944--1955. At the turn of the 19th to 20st and 20th to 21st centuries the solar corona shape and, possibly, the global magnetic field correspond to the configuration close to the octopole one (see Fig. 8).

The period of variation in the corona's shape during the epoch of minimal activity is about $100 \div 120$ years (see Fig. 4), which is close to the Gleissberg cycle for the sunspots, but probably precedes it to some extent in phase (*Hathaway*, 2010, Fig. 34).. The maximum of the secular variation in the global magnetic field of the Sun occurred before cycle 19 and preceded the sunspot activity maximum. This allows us to put forward the hypothesis that secular variations in the solar activity are caused by secular modulation of the global magnetic field of the Sun. Another conclusion of this work is the supposition that the slowly changing component of the geomagnetic activity derived from the data on the *aa* index from changes in the dipole component of the large-scale field of the Sun (Figure 7).

The link between the coronal shape, geomagnetic disturbances and solar wind parameters can be clearly seen in 11-year activity cycle. Global magnetic field can lead to non-radial spreading of the coronal streamers and probably, to the spreading of the solar wind. In minimal activity when the value of the global field reaches its maximum, coronal rays are deflected aside from the helio-equator. Here with lines of force influence the elapsing flux of the solar wind, squeezing it towards the ecliptic plane (Fig. 19). It is possible that long-term changes of geomagnetic disturbances that do not tend to zero even in the minimum activity are not conditioned this very deflection of the solar wind.

The awareness of the variations of the angle $\Delta\theta$ with the phase of a cycle is important for theoretical models describing the structure of the corona and the geometry of the magnetic field above the solar limb (Wang, 1996). The applied aspect of the studies for the deviations of coronal streamers, along with the plasma flows in the solar wind, is also of great importance, since the deviation may affect the geo-efficiency of the solar wind impact. The presence of long-term trends in the solar corona structure can be caused by changes in the configuration of the global magnetic field of the Sun. The non-radial propagation of the solar wind may explain the relationship between geomagnetic indices detected during the minima of activity and the amplitude of a subsequent solar cycle (Ohl, 1966). Indeed, the relation between the amplitude of the angles $\Delta\theta$ at the minima of the 22-th – 24-th cycles is close to that for the index of the large-scale magnetic field of the Sun in these times (Tlatov, 2009), and this field is one of prognostic indices of the solar activity. Thereby, we may suggest a link between the large-scale magnetic field and the deviation of coronal rays towards the equator, which in turn affects the level of geomagnetic indices at a solar minimum as well as solar-terrestrial relations.

The deviation coronal streamers $\Delta\theta$ occur as the cyclic process appreciable at all solar latitudes (Figure 13,14). Probably, it is connected with global processes, for example hemispheric current layer (Filippov, 2009).

In the analysis we have considered the various structures, including possibly producing a unipolar coronal streamer belt (called' 'chains of streamers'' or ''streamer belt without a neutral line'' by (Eselevich et al., 1999) as well as the bipolar streamer belt (Zaho & Webb, 2003). These streamers vary in brightness and location of footpoint. Our analysis of separation of their brightness and latitude showed that cyclic changes in the nonradial parameter $\Delta\theta$ exist for various types of coronal streamers. The good agreement between the change in time for parameter $\Delta\theta$ change the angle of inclination of HCS, suggest that the parameter of nonradiality, is possibly conditioned by the existence of current in the heliospheric current layer. The current in the HCS can be estimated from the parameter $\Delta\theta$.

In the epoch of the 18-19 activity cycles in the middle of the last century the shape of the solar corona in its minimal activity mostly corresponded to the corona of the minimal type, that is – coronal radiance spreaded along helio-equator (Fig. 3). Just in that period the highest level of geomagnetic activity was observed (Fig. 6), what is more, not only during the years of solar spots maximum, but also in the minimal activity. We can assume that during that period global magnetic field of the Sun most effectively squeezed the solar wind flux towards the heio-equator plane, which led to such an effect.

Variation changes of the low and high harmonics order are illustrated on figure 20. Here the amplitudes of the harmonics were smoothed by the "running window" of 11 years long. One introduces amplitudes of axisymmetric mode for dipole and quadruple (l=1,2; m=0) and for harmonics, characterizing sector structure of the higher harmonics (l>3; m>0). One can notice that changes in time within the amplitudes of these two types occur in antiphase. One should also notice that axisymmetric modes (m=0) show that magnetic structures of the opposite polarities are situated on different sides of the equator. On the contrary, the amplitude growth of the sector harmonics shows that opposite signs structures are drawn along meridians. The amplitudes of the axisymmetric harmonics reach their maximum in

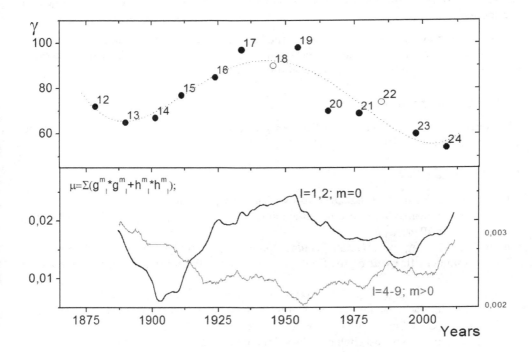

Fig. 20. Variation changes of the low and high harmonics order are illustrated on figure 20. Here the amplitudes of the harmonics were smoothed by the "running window" of 11 years long. Bottom panel introduces amplitudes of axisymmetric mode for dipole and quadruple (l=1,2; m=0) and for harmonics, characterizing sector structure of the higher harmonics (l>3; m>0). At the top changes of parameter γ form the solar corona.

the epoch of years 1940-1950, while sector harmonics were minimal at that moment. Such allocation corresponds to the Babcock hypothesis concerning magnetic fields generation. In fact, reconnection of the opposite through equator fields causes mutual destruction of the fields of the leading polarity, and it should cause the solar activity growth. But on the contrary, reconnection of the magnetic field along the meridians does not promote the creation of global field of dipole type and leads to the decline of solar activity level.

Consequently, in the epoch of big solar activity cycles the growth of geomagnetic activity takes place not only due to the result of active processes increase, but also under the influence of solar wind's formation by the global magnetic field of the Sun in activity minimums. The following fact contributes to this – before the high cycles of activity the intensity of global magnetic field is also high (Fig. 5). It is possible that age-long cycle of changes of solar wind parameter and geomagnetic activity, close to Gleissberg cycle is caused not only by the level of activity, but also by long-term changes of the large-scale magnetic field.

Thereby, we may suggest a link between the large-scale magnetic field and the deviation of coronal rays towards the equator, which in turn affects the level of geomagnetic indices at a solar minimum as well as solar-terrestrial relations.

5. Acknowledgements

This paper was supported by the Russian Fund of Basic Researches and Program of the Russian Academy of Science.

6. References

Arge, C. N., & Pizzo, V. J. (2000). Improvement in the prediction of solar wind conditions using near-real time solar magnetic field updates, *J. Geophys. Res.*,Vol. 105,pp. 10465-10480.

Cliver, E. W. & Ling, A. G. (2002). Secular change in geomagnetic indices and the solar open magnetic flux during the first half of the twentieth century, *J. Geophys. Res.*, SSH 11-1, 107.

Eselevich, V. G., & Eselevich, M. V. (2002). Study of the nonradial directional property of the rays of the streamer belt and chains in the solar corona, *Solar Phys.*, Vol. 208, pp.

Eselevich, V. G., Fainshtein, V. G. & Rudenko, G. V., (1999). Study of the structure of streamer belts and chains in the solar corona, *Solar Phys.* , Vol. 188,pp. 277-297.

Filippov, B. P. (2009), Non-radial coronal streamers in the course of the solar cycle, *Astronomy Reports*, Vol. 53,pp. 564-568.

Ganskiy, A. P. (1897). *Die totale sonnenfinsterniss am 8. august 1896*, Proc. Royal. Akad. Sci. (in Russian), Vol. 6, 251.

Golub, L. & Pasachoff, J. M. (2009). *The Solar Corona, second edition*, (Cambridge University Press).

Gulyaev, R. A. (1998). New results of observations of eclipsing the solar corona, *in Proceedings of the Conference "A New Cycle of Solar Activity: Observations and Theoretical Aspects" (in Russian)*, ed. A. V. Stepanov, (Pulkovo, St.Petersburg), pp. 61-66.

Hathaway, D. H. (2010). The Solar Cycle, *Living Rev. Solar Phys.*, 7, (http://solarscience.msfc.nasa.gov).

Hoeksema, J. T. & Scherrer, P. H. (1986). An atlas of photospheric magnetic field observations and computed coronal magnetic fields: 1976-1985, *Sol.Phys.*, Vol. 105,pp. 205-211.

Jian, L. K.; Russell, C. T. & Luhmann, J. G. (2011). Comparing Solar Minimum 23/24 with Historical Solar Wind Records at 1 AU, *Solar Phys.*, Doi 10.1007/s11207-011-9737-2.

Loucif, M. L. & Koutchmy S. (1989). Solar cycle variations of coronal structures, *Astronomy and Astrophysics Supplement Series*, Vol. 77, pp. 45-66.

Makarov, V. I. & Sivaraman, K.R. (1989). Evolution of latitude zonal structure of the large-scale magnetic field in solar cycles, *Solar Phys.*, 119, 35-44.

Makarov, V. I. & Tlatov A. G. (2000). The Large-Scale Solar Magnetic Field and 11-Year Activity Cycles, *Astron. Rep.*, Vol. 44, pp. 759-764.

Meyer-Vernet N. (2007). *Basics of the Solar Wind*, ISBN 978-052-1814-20 Cambridge University Press

Naegamvala, K. D. (1902). Report on the total solar eclipse of January 21-22, 1898, (Bombay), 49.

Nikolskiy, G. M. (1955). Forecast form the solar corona on June 20, 1955, *Astron. Circular*, 160, 11-12.

Ohl, A. I. (1966). *Soln. Danie*, N 12, 84.

Parker E. (1958). Dynamics of the Interplanetary Gas and Magnetic Fields, *The Astrophysical Journal*, Vol. 128, pp. 664-676

Pasachoff, J. M., Ružin, V., Druckmüller, M., Druckmülerová, H., Belik, M., Saniga, M., Minarovjech, M., Marková, E., Babcock, B. A., Souza, S. P. & Levitt, J. S. (2008). Polar Plume Brightening During the 2006 March 29 Total Eclipse, *ApJ*, Vol. 682, pp.

Pneuman, G. W. & Kopp, Roger A. (1971). Gas-Magnetic Field Interactions in the Solar Corona , *Solar Phys.*, Vol. 18, pp.258-270

Saito, T., Shibata, K., Dere, K.P. & Numazawa, S. (2000). Non-Radial Unipolar Coronal Streamers in Magnetically High Latitudes and Radial Bipolar Streamers at the Magnetic Equator of the Sun, *Adv. Space Res.*, Vol. 26,pp. 807-810.

Tlatov, A. G. (2009). The Minimum Activity Epoch as a Precursor of the Solar Activity, *Solar Phys.*, Vol. 260, pp. 465-477.

Tlatov, A. G. (2010a). The centenary variations in the solar corona shape in accordance with the observations during the minimal activity epoch, *Astronomy and Astrophysics*, Vol. 522, id.A27

Tlatov, A. G. (2010b). The Non-radial Propagation of Coronal Streamers within a Solar Cycle, *ApJ*, Vol. 714,pp. 805-809.

Vasil'eva, V. V. (1998). Recovery of large-scale synoptic maps of magnetic fields for the period 1880-1894, *in Proceedings of the Conference "A New Cycle of Solar Activity: Observations and Theoretical Aspects" (in Russian)*, ed. A. V. Stepanov, (Pulkovo, St.Petersburg), 213-216.

Vsekhsvyatskiy, S. K., G. M. Nikolskiy, V. I. Ivanchuk, A. T. Nesmyanovich, et al. (1965). *Solar Corona and Corpuscular Radiation in the Interplanetary Space*, (Kiev, Naukova Dumka), 293.

Vsekhsvyatskiy, S. K., G. M. Nikolskiy, V. I. Ponamarev E.A. & Cherednichenko V. I. (1955). On the corpuscular radiation from the sun, *Astronomy Rep.*, Vol. 32, pp. 165-176

Waldmeier, M. (1976). *Astron. Mitt. Eidgen. Sternw. Zurich*, Vol. 351,pp. 13.

Wang Y.-M. (1996). Nonradial Coronal Streamers, *ApJ*, Vol. 456, pp. L119-L121.

Wang, Y.-M., & Sheeley, N. R., Jr. (1990). Solar wind speed and coronal flux-tube expansion, *ApJ*, Vol. 355, pp. 726-732

Zhao, X. P. & Webb, D. F. (2003). Source regions and storm effectiveness of frontside full halo coronal mass ejections *JGR*, Vol. 108, pp. SSH 4-1.

2

Variability of Low Energy Cosmic Rays Near Earth

Karel Kudela
Institute of Experimental Physics,
Slovak Academy of Sciences, Kosice
Slovakia

1. Introduction

Almost a century ago Victor Hess in his balloon experiments discovered cosmic rays (Hess, 1912). The history in understanding the nature, physical mechanisms affecting the temporal and spatial variability of cosmic rays (CR) observed on the ground and within the atmosphere, as well as of the anisotropy of secondary particles due to geomagnetic effects and due to the state of atmosphere is reviewed e.g. in the monographs (Hillas, 1972; Dorman, 2004; 2009; Rossi and Olbert, 1970 and references therein). Progress in clarifying mechanisms controlling the secondary CR flux required the design and construction of new detectors. Production of secondary CRs is reviewed in the books (Grieder, 2001; 2010). Until half of the past century the elementary particle physics was driven mainly by CR studies. With the improvement of the acceleration technique, the domain of CR for subnuclear physics research, was shifted to high energies. Even today the high energy CRs observed by the large ground based detectors remain the only one source of information about the particles of extremal energy. The astroparticle physics aspect of CR is described e.g. in the books (Grupen, 2005; Gaisser, 1990; Hayakawa, 1969). Relations of high energy astrophysics to CR physics include the book (Longair, 1981 and the newer versions).

The important step in CR studies represented the International Geophysical Year in 1957-1958. The new detectors and its networks were built (Simpson 1958; Hatton, 1971; Stoker, 2009 among others) and CR research attracted many scientists, engineers and students. Most important stimulus for the progress in low energy cosmic rays (LECR) was the launch of the first satellites of Earth which meant beginning of the Space Era. Shortly after measurements of radiation on the first satellites, the new populations of energetic particles in space (not providing secondary „signal" on the Earth) have been found. That direction of research was important also for plasma physics. Processes of particle acceleration, transport and losses in the plasma regions in space, where conditions are inimitable for laboratory plasma experiments, brought new knowledge into plasma physics. There are several books and papers summarizing/illustrating in detail CR physics in its history. At low energies to which this chapter is devoted, much more informations can be found e.g. in (Dorman, 1974; Bieber et al., 2001; Vainio et al., 2009; rapporteur papers in the proceedings of ICRCs). Here we mention selected results in the experimental studies related to LECR with empasis to the papers published in the past 2-3 years.

2. LECR variations

Two populations of LECR can be assigned, namely (1) at the energies above, and (2) below the atmospheric threshold (~ 400 MeV for protons). Above that energy the ground based measurements provide the direct information about changes in the flux of primaries. For lower energies the measurements of particles on satellites, space probes, rockets and balloons are most relevant. Two types of experimental devices on the ground are very important for the detection of LECR flux variability above the atmospheric threshold, namely neutron monitors (NM) and muon telescopes (MT) sensitive to different ranges of energy spectra of primaries.

Galactic CRs entering the heliosphere are affected by the interplanetary magnetic field (IMF) which, especially in the inner heliosphere, is controlled by the solar wind plasma having higher energy density than IMF. Concept of field lines frozen in the high conductivity solar wind plasma is often used. CR in the inner heliosphere having lower energy density than the IMF, can be assumed as a specific "autonomous" population of particles. In the outer heliosphere, however, the relations are changing. The heliosphere via the IMF is modulating the CR flux at the low edge of its energy spectra. In addition, it contributes itself to energetic particle populations by acceleration at the Sun as well as at plasma discontinuities in the interplanetary space, and within the magnetospheres of planets with strong magnetic field. Additionally, heliosphere is transparent for access of neutral atoms from outer space. They can be ionized in the heliosphere and subsequently accelerated, which contributes to the suprathermal particle population known as anomalous cosmic rays (Garcia-Munoz et al., 1973). Modulation of CR depends on its primary energy. Solar wind with the embedded IMF flowing outward from the Sun screens the access of primary CR into the heliosphere. The physical framework in which the energetic particles and CR propagate in the heliosphere is also denoted as heliospheric magnetic field, HMS (recent review e.g. by Balogh and Erdös, 2011). Below few hundreds MeV practically no galactic CR enter the inner heliosphere (Jokipii, 1998). The modulation below ~ 10 GeV is present even during solar activity minimum. The main feature of a long term variation of low energy galactic CR (GCR) near Earth is the anticorrelation of the flux with solar activity having about 11 year cyclicity. In addition to ~11 year periodicity, the ~ 22 year modulation cycle in CR flux due to solar magnetic field polarity reversals, is observed (e.g. Webber and Lockwood, 1988). Around the epoch of solar activity maxima there is observed a double-peak structure in many solar activity factors. A distinctive minimum between the peaks is called Gnevyshev gap (Storini et al., 2003 and references therein). In CR this minimum is connected e.g. with decrease of ~27 day quasi-periodicity.

Theory of CR transport, used basically until present with several small modifications, was described first by E.N. Parker (1965). Energetic particles in the interplanetary space walk randomly in irregularities of the large-scale IMF when irregularities are moving with the solar wind velocity. The distribution function determined by a Fokker-Planck equation describes the time evolution of the probability density function of the position and momentum of particle. In addition to the convection and diffusion, CRs experience two additional effects. One of them is the acceleration or deceleration. Solar wind plasma is expanding in free space and compressing at the shocks near the planets or in the interplanetary space. The inhomogenities with different IMF become mutually more distant or more close to each other. This leads to the adiabatic cooling or heating due to multiple interactions of CR with inhomogenities. Modulation of GCR in the inner heliosphere is controlled by convection in solar wind, diffusion, particle drifts and adiabatic energy losses. Using the experimental data from various space missions, an overview of the current understanding of the modulation over the past decade is given by (Potgieter, 2011). At low energies, 10–100 MeV, where the adiabatic cooling plays the primary role, its effect can be best seen in the framework of the force-field approximation (Gleeson & Axford 1968; Fisk et al. 1973). In addition, the curvature and gradB drifts in IMF play role in the modulation. Gyration of particle around the field line is faster than scattering. Thus particles are subject of drift due to large scale spatial structure of the IMF. CR transport and modulation was studied in many papers (e.g. Jokipii, 1966; 1967; Potgieter and Le Roux, 1994; Zank et al., 1998; Cane et al., 1999 among others) and it is examined further along with the new knowledge of CR flux at various points in the heliosphere and near its boundaries. For the transport equation the determination of the coefficients deduced from the measurements at various positions and energies is important. Reviews of transport coefficients for LECR is e.g. in (Palmer, 1982; Valdés-Galicia, 1993). Recently (Kecskeméty et al., 2011) examined the energy spectra of 0.3–100 MeV protons and found, that at the lower energies, the galactic particle spectra are significantly steeper than the $J(E) \sim E$ one, predicted by analytical approximations, such as the force-field model of modulation. Usoskin et al. (2005) provide a long series of a parameter allowing for a quantitative estimate of the average monthly differential energy spectrum of CR near the Earth for long time interval.

CR, measured directly by its secondaries at Earth over more than half of century, indicate complicated structure in its temporal behavior. The nucleonic component of primary CR is appropriate to study solar modulation from ground based measurements (e.g. Storini, 1990). CR modulation as observed from the Earth is discussed and reviewed e.g. by (Belov, 2000). Single point measurements are influenced by the large scale structure of IMF over the heliosphere, which, in the given time, in addition to the measured IMF and solar wind e.g. at L1, includes also "memory effect" of solar activity due to relatively slow outward motion of solar wind with the IMF inhomogenieties, if compared to the speed of CR particles. Long term modulation is observed as a series of steplike decreases. Outward propagating diffusion barriers were identified as merged interaction regions, MIRs (Burlaga et al., 1985). A comprehensive review of various CR intensity variations over different time scales that have been conducted over 1970s and 1980s can be found in paper (Venkatesan and Badruddin, 1990) and in references therein. Recently (Strauss et al., 2011) review some of the most prominent CR observations made near Earth, and indicate how these observations can be modelled and what main insights are gained from the modelling approach. Also, discussion on drifts, as one of the main modulation processes, is given as well as how the drift effects manifest in near Earth observations. Specifically, discussion on explanation of the observations during past unusual solar minimum, is included. Siluszyk et al. (2011) developed a two-dimensional (2-D) time dependent model of the long period variation of GCR intensity, where they included the slope of power spectrum density of IMF fluctuations, magnitude of B, tilt angle of heliocentric current sheet and effect of particle drift depending on solar magnetic field polarity. Solar modulation of GCR over the past solar cycles is described and discussed in detail e.g. by (Chowdhury et al., 2011). Recent solar minimum was specific one, unusually long and deep (e.g. Badruddin, 2011). The modulation of CR was minimal for the more than 70-year-long period of direct measurements (Bazilevskaya et al., 2011). In 2009 (Stozhkov et al., 2011) recorded the highest

CR fluxes (particles with energy > 0.2 GeV) in the history of the CR measurements in the stratosphere. An increase of the flux on NMs during that minima was also reported (e.g. Moraal and Stoker, 2010). GCR modulation for solar cycle 24 began at Earth's orbit in January 2010 (Ahluwalia and Ygbuhay, 2011). That paper reports that some NMs are undergoing long-term drifts of unknown origin. Such effects have to be examined in detail because of correct understanding the long term CR variations from the direct measurements. Corrections of data in the NM network are discussed also e.g. by (Dvornikov and Sdobnov, 2008). There are several sources of the data from CR measurements by NMs (recently e.g. NMDB data base at http://nmdb.eu).

Modulation of the low energy component of GCR inside the heliosphere gives us insight on the relevance of solar phenomena that determine the structure and evolution of the heliosphere. Reviews of the advancement in the comprehension of the phenomena controlling the transport of LECR in heliosphere can be found e.g. in (Valdés-Galicia, 2005). Time profiles of CR observed at a given position of NM on the ground are result of superposition of many transitional effects due to the temporary and spatially changing structures of IMF irregularities within the heliosphere, by the rotation of the Earth with detector, by the effects of magnetosphere and by the variable state of the atmosphere. In addition to the network of NMs, the information about the variability of CR at higher energies provide the detectors in relatively new installations. The decription of few of such installations along with the first results are reported e.g. in papers (Augusto et al., 2011; De Mendonca et al., 2011; Maghrabi et al., 2011).

Charge dependent modulation is important point in the CR drift models (e.g. Kóta and Jokipii, 1983; Potgieter and Moraal, 1985). The drift effects have to be marked differently for CR particles with the opposite sign of electric charges during epochs with opposite polarities of the solar magnetic field. Thus the interest to the measurements of positrons and antiprotons is increasing. Important finding was done by PAMELA experiment (Adriani et al., 2009; 2010). The authors present data on the positron abundance in the CR in the energy range 1.5 - 100 GeV with high statistics. The data deviate significantly from predictions of secondary production models, and the authors stress that it may constitute the first indirect evidence of dark matter particle annihilations, or the first observation of positron production from near-by pulsars. The evidence of solar activity affecting the abundance of positrons at low energies is reported too. This result was recently confirmed independently by the observations of FERMI (Ackermann et al., 2011), where the authors report first time measurement of absolute CR positron spectrum above 50 GeV, and indicate that the fraction has been determined above 100 GeV. Increase of positron fraction with energy between 20 and 200 GeV is found. The future measurements with greater sensitivity and energy resolution, such as those by AMS-02, are necessary to distinguish between many possible explanations of this increase. Serpico (2011) summarizes the global picture emerging from the data and recapitulates the main features of different types of explanations proposed. Testing of different scenarios and inferring some astrophysical diagnostics from current/near future experiments is also discussed.

2.1 Irregular variations

Coronal Mass Ejections (CME) are causing changes in CR intensity measured at Earth. Although the decrease in CR coincident with geomagnetic field depression (horizontal component) was first observed by Forbush already in 1937 (in the book Van Allen, 1993, p. 117), this phenomena attracts the CR physicists until present. Forbush decreases are generally correlated with co-rotating interaction regions (CIRs) or with the Earth-directed CMEs from the Sun (e.g. Prasad Subramanian, 2009). The characteristics of CMEs in the inner heliosphere are discussed in detail by (Gopalswamy, 2004). One of the first papers suggesting that the solar cycle dependent modulation of GCR can be explained by CMEs and by related IMF inhomegeneities in the heliosphere was that by (Newkirk et al., 1981). Correlation of CR intensity and CME ocurrence at a single NM with high geomagnetic cut-off is reported e.g. by Mishra et al. (2011) for different solar magnetic field polarities.

The Forbush decrease (FD) or cosmic ray storm is produced as a result of the transient diffusion-convection of CR caused by the passage of IMF shock wave. Nagashima et al. (1992) show that the storm is frequently accompanied by non-diffusion-convection-type phenomena, depending on local time, as precursory decrease of CR at the front of the shock in the morning hours of local time, and a post-shock increase different from the diurnal variation. FDs are analyzed by global spectrographic method using worldwide network of NMs (e.g. Sdobnov, 2011 and references therein). Recently (Dumbovic et al., 2011) performed statistical study of the relationship between characteristics of solar wind disturbances, caused by interplanetary CMEs (ICMEs) and corotating interaction regions, as well as with properties of FDs. It was found that the amplitudes of the CR depression are primarily influenced by the increase in magnetic field strength and fluctuations, and the recovery phase also depends on the magnetic field strength and size of the disturbance. The use of FDs for space weather studies is discussed by (Chauhan et al., 2011). For this application the complexity of relations between the geomagnetic storms and FDs must be assumed. Mustajab and Badruddin (2011) critically analyze the differences in geoeffectiveness due to different structures and features, with distinct plasma/field characteristics. Distinct relations of FDs and geomagnetic activity measured by Dst in different events is reported by (Kudela and Brenkus, 2004; Kane, 2010). Richardson and Cane (2011) analyzed large number of ICMEs and their associated shocks passing the Earth during years 1995 – 2009. They found that magnetic clouds are more likely to participate in the deepest GCR decreases than ICMEs that are not magnetic clouds. Examining simultaneous observations of FD events by different CR stations remains a subject of interest. Variability in the manifestations of FDs demonstrates that there are still open questions in this field (e.g. Okike and Collier, 2011; Pintér et al., 2011). Study (Oh and Yi, 2009) may support the hypothesis that the simultaneous FDs occur when stronger magnetic barriers pass by the Earth, and in contrast that the nonsimultaneous FDs occur only if the less strong magnetic barriers pass the Earth on the dusk side of the magnetosphere. The FDs are observed not only by NMs but also at higher energies of primary CRs (e,g. Braun et al., 2009; Abbrescia et al., 2011; Bertou, 2011) or by a lead free NM (Mufti et al., 2011). Rigidity spectra of FDs and their relations to the index of power spectra density (PSD) of IMF fluctuations in the frequency range in which the interaction of IMF with CR can be efficient, is studied in papers (Alania and Wawrzynczak, 2008; 2011). The theoretically derived relationship between rigidity spectrum exponent (γ) and exponent (ν) of IMF PSD is confirmed. Figure 1 illustrates different relations of FDs to geomagnetic activity.

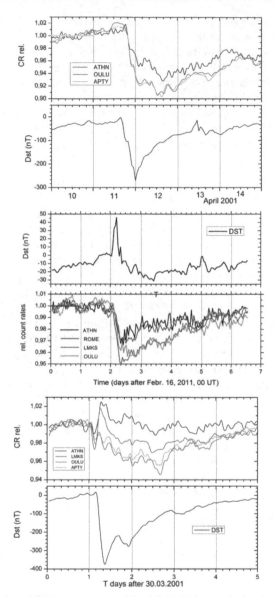

Fig. 1. Three decreases of GCR as measured by NMs (Athens, Oulu, Apatity, Rome, Lomnicky stit) with different relations to geomagnetic activity. Upper panel: CR storm accompanied by Dst depression. FD is better pronounced at low cut-off rigidity NMs (Apatity, Oulu) than at higher one (Athens). Middle panel: FD observed at several european NMs not accompanied by any Dst depression (IMF Bz>0). Lower panel: At low cut-off NMs CR decrease is seen (geomagnetic cut-offs near or below the atmospheric ones), while at middle and high cut-off positions (Lomnicky stit, Athens) an increase is seen due to the improvement of magnetospheric transmissivity.

Analysis of FDs is important for better understanding of the magnetic field structure related to shock waves and fast streams originating at the Sun. Quenby et al. (2008) examined the temporal history of the integral GCR fluence (≥100 MeV) measured by the high-sensitivity telescope (HIST) aboard the Polar spacecraft, along with the solar wind magnetic field and plasma data from the ACE spacecraft during a 40-day period encompassing September 25, 1998 FD. The authors also analyzed FD and energetic storm particle event on October 28, 2003, one of the largests in the past decades. Short-scale GCR depressions during a test period in September through October 1998 did not show correlation with changes in magnetic scattering power or fluctuations in solar wind speed or plasma density. However, IMF and solar wind data during the test period of FD suggest the presence of ICME. Mulligan et al. (2009), using the high resolution energetic particle data from ACE SIS, the Polar high-sensitivity telescope, and INTEGRAL's Ge detector measuring GCR background with a threshold of 200 MeV, show similar, short-period GCR variations in and around the FD. NMs have lower statistics. Earlier paper (Kudela et al., 1995) indicated that the Dst decreases are correlated with the „prehistory" of CR fluctuations on time scales longer than tens of minutes especially during the years with high solar activity. In the future high temporal resolution data on CR are needed, and the analysis based on combination of NM data with the satellite measurements by detectors having large geometrical factors is important (e.g. Grimani et al., 2011).

Another class of irregular variations of CR intensity observed in the vicinity of Earth are high energy particle populations accelerated in the solar flares or at the discontinuities in interplanetary space. Two types of solar energetic particle (SEP) events, namely impulsive and gradual ones, have been recognized since 1980s (e.g. Lin, 1987). Reames (1999) summarized the knowledge about energetic particle populations coming from solar flares, from shock waves driven outward by CMEs, from magnetospheres of the planets and bow shocks and reviewed various acceleration processes throughout the heliosphere. Miroshnichenko (2001) surveyed in detail the results of solar CR investigations since 1942, with including a large amount of data, obtained during long time period of observations of SEP. The book also covers theoretical models and gives an extensive bibliography. Recently Hudson (2011) reviews the knowledge of solar flares with the focus on their global properties. Flare radiation and CME kinetic energy can have comparable magnitudes, of order 10^{25} J each for an X-class event, with the bulk of the radiant energy in the visible-UV continuum. The author argues that the impulsive phase of the flare dominates the energetics of all of these manifestations, and also points out that energy and momentum in this phase largely reside in the electromagnetic field, not in the observable plasma. Barnard and Lockwood (2011) constructed a database of gradual SEP events for 1976-2006 using mainly data of > 60 MeV protons. Although number of events decreases when solar activity is low, the events during solar minimum are observed with higher fluence. Thus, very strong flares may be more likely at lower solar activity. Ground level events (GLE) are observed by NMs when the energy of accelerated particles in the flare or in interplanetary space exceeds the atmospheric threshold and geomagnetic cut-off rigidity. Characteristics of GLEs for the past solar cycles are summarized and discussed e.g. by (Gopalswamy et al., 2010; Andriopoulou et al., 2011). Moraal and McCracken (2011) analyzed all GLEs for the cycle 23. Three of the 16 GLEs have a double-pulse structure. They are associated with western flares and have good magnetic connection to the Earth. All have fast anisotropic first pulse followed by a smaller, gradual, less anisotropic second pulse. Vashenyuk et al. (2011) present a GLE modeling technique applied for 35 large GLEs for the period 1956 – 2006 and obtained features of prompt and delayed components of relativistic solar particles. Kurt et al. (2011) studied signatures of protons with energy above several hundred of MeV associated with major solar flares and observed by NMs during GLEs. The authors revealed that the delay of the earliest arrival time of high-energy protons at 1 AU with respect to the observed peak time of the solar bursts did not exceed 8 min in 28 events. This indicates that efficient acceleration of protons responsible for the GLE onset is close to the time of the main flare energy release. For the GLE observations are important high altitude NMs due to their high count rate and high statistics. List of GLEs observed at one high mountain NM can be found in (Kudela and Langer, 2008).

Important source of the information about protons accelerated near the Sun and about their interactions with residual solar atmosphere are solar gamma-rays and neutrons. Their observation near Earth is not affected by magnetic field as it is in the case of protons GLEs, SEP events. Production of γ -rays and neutrons results from convolution of the nuclear cross-sections with the ion distribution functions in the atmosphere. Recently Vilmer et al. (2011) reviewed the γ-ray and neutron observations with the emphasis on the very detailed RHESSI measurements, namely the high spectral resolution revealing line shapes and fluences, and gamma-ray imaging technique. The authors point out also still open question for the study of high energy neutral emissions from the Sun. Chupp et al. (1973) reported first observations of gamma ray lines from solar flares in August 1972 using data from OSO-7 satellite. Ramaty et al. (1979) reviewed the gamma-ray line emission from the Sun due to nuclear deexcitation of ambient nuclei following the interactions of accelerated particles. Although Biermann et al. (1951) long time ago proposed that high energy neutrons created by nuclear interactions of protons with the atmosphere of the Sun can be observed on Earth's orbit, first direct indication of solar neutrons on the ground was reported from high mountain NMs Jungfraujoch and Lomnický štít in the flare event June 3, 1982 (Debrunner et al., 1983; Efimov et al., 1983) and on satellite by electrons from the neutron decay (Evenson et al., 1983). After that event the interest to detection of solar neutrons increased. Several NMs started to measure with better temporal resolution, and new experimental devices for detection of gamma rays and neutrons both on the ground as well as on the satellites were constructed. High energy gamma rays and neutrons during several solar flares have been observed in the past decade also e.g. on low altitude polar orbiting satellite Coronas-F (Kuznetsov et al., 2006; 2011; Kurt et al., 2010). Review on experimental and theoretical works related to solar neutrons is in (Dorman, 2010a).

2.2 Periodic and quasi-periodic variations

The quasi-periodic and periodic variations in CR intensity observed at Earth are studied for rather long time. The solar diurnal wave, being the fixed one, was checked starting from papers (Forbush, 1937; Singer, 1952; Thompson, 1938; Brunberg and Dattner, 1954; Ahluwalia and Dessler, 1962). Studies of its higher harmonics, namely of semi-diurnal one, can be found in papers since (Ahluwalia, 1962; Nicolson and Sarabhai, 1948) and on the tri-diurnal starting probably from (Mori et al., 1971; Ahluwalia and Singh, 1973). At longer quasi-periodicities first attention was paid to ~ 11 yr, ~22 yr and ~27 d variabilities. The detailed review is in the books (Dorman, 1975; 2004) and references therein.

For the shape of the PSD we examined the long time series of Climax NM, skipping the data influenced by the GLEs, interpolating linearly the gaps, and applying the FFT method. PSD obtained is plotted in Figure 2.

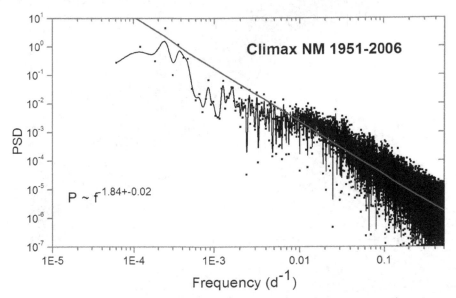

Fig. 2. The power spectrum density (PSD by FFT method) of daily averages of count rate of Climax NM (dots). Data downloaded from ftp://ulysses.sr.unh.edu/NeutronMonitor/DailyAverages.1951-.txt. Line is for the cubic spline connection.

The spectral density at higher frequency ranges is plotted in Figure 3.

At frequencies $f < 5.8 \times 10^{-6}$ Hz (upper panel of Figure 3) the slope is consistent with the theory (Jokipii and Owens, 1974b) where the authors indicate that including the effects of non-field-aligned diffusion, which dominates the power spectrum of NMs at low frequencies ($< 5 \times 10^{-6}$ Hz) produces a spectrum of f^{-2}. The lower panel clearly indicates the presence of the fixed frequencies of the diurnal, semi-diurnal and tri-diurnal waves. Probably the fourth harmonic is present too. Index of the spectra is lower, around -1.5. At $f > 5.10^{-6}$ Hz the theory with field-aligned diffusion is satisfactory for explanation of the shape (Jokipii and Owens, 1974a).

The solar diurnal anisotropy fixed at PSD as a single periodicity is resulting from the co-rotational streaming of particles past Earth (Duldig, 2001). Kudela et al. (2011) indicate the difference in the slopes of PSD at NM Lomnický štít (data at http://neutronmonitor.ta3.sk) for different phases of a solar activity cycle, namely the hardening of the PSD at solar minimum in comparison with the maximum. Similar behavior is found in the spectral analysis of the IMF magnitude B. Such picture is in qualitative agreement with the slopes of IMF measured in different solar cycle phases at large distances (Burlaga and Ness, 1998). El-Borie and Al Thoyaib (2002) studied power spectra of CR in the range 2 – 500 days and found significant differences in the individual spectra of solar maxima for different cycles.

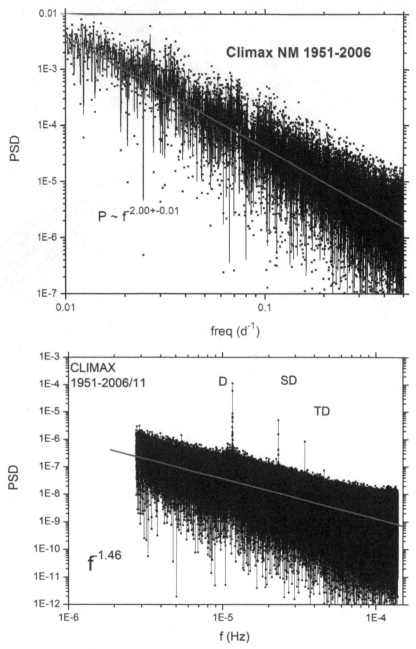

Fig. 3. Power spectrum density of Climax NM data at two ranges of frequencies. For the upper panel the daily means were used, for the lower one the hourly corrected ones. D, SD and TD is foor diurnal, semi-diurnal and tri-diurnal variation. While the amplitude of D has ~11 yr variation, the phase has ~22 yr one.

Spectra in range 2.7 x 10^{-7} – 1.4 x 10^{-4} Hz measured by muon underground detector have been examined by (Sabbah and Duldig, 2007). Flatter spectra having lower power when the interplanetary magnetic field (IMF) is directed away from the Sun above the heliosphercic current sheet (A > 0) than when the IMF is directed toward the Sun above the current sheet (A < 0), are reported.

For the description of the ~27 day variability related to the solar rotation, the daily count rate means of NM Climax was filtered in the frequency range corresponding to the time scale 25 – 33.3 days and the wavelet transform was applied on the data (Figure 4).

The double-peak structure, namely with maxima at ~ 27 days and another at ~ 31 days is found around solar activity maxima, similar to (Dunzlaff et al., 2008) based on data of GCR, EPHIN on SOHO. The structure is more complex during the long time period. Transport models (Gil et al., 2005) and measurements analyzed in paper (Richardson, 2004) suggest the dependence on solar magnetic field polarity. Vivek Gupta and Badruddin, (2009) found that the average behavior of GCR-oscillations during Carrington rotation is different in A > 0 from that in A < 0 epoch. Correlation of solar wind speed with GCR intensity during the course of Carrington rotation is stronger for A > 0 than for A < 0. The amplitudes of GCR-oscillations show somewhat weak dependence on the tilt angle of the heliospheric current sheet. Krymsky et al. (2008) indicate that temporal change of the power spectrum of ~13.5-day and ~27-day variations repeats the power spectrum change of the number of sunspots and tilt angle of the current sheet, and that the dependence of ~ 27-day variation on the polarity of general magnetic field of the Sun is not found. This feature has to be examined in future by wavelet technique in more detail. The ~27 day variation correlates with B, Bz, v, and B(v x B) – (Agarwal et al., 2011a). Similarly to ~27-day variability we examined the vicinity of ~13.5 and ~9 day periodicity contributions. At ~13.5 days we confirm the result by (Filisetti and Mussino, 1982) using ionisation chamber data, indicating the maximum contribution is correlated with the sunspot number. At ~9 days the results can be found in paper (Sabbah and Kudela, 2011).

Rieger et al. (1984) reported 154 day periodicity in solar X-ray and gamma ray flares. Pap et al. (1990) examined various periodicities in solar activity time series. There is no explanation for 150-157 day period found in several data sets. 154-day periodicities in the near-Earth IMF strength, in solar wind speed (Cane et al., 1998) and in solar proton events (Gabriel et al., 1990) have been reported. Hill et al. (2001) using Voyager 1 data, have shown that at anomalous CRs the quasi-periodic variations are in phase, with O, He having periods ~ 151 days, while protons exhibit a period ~ 146 days. Results about quasi-periodicity ~150 days in CR measured on the ground can be found e.g. in papers (Mavromichalaki et al., 2003; Kudela et al., 2010).

Another quasi-periodicity observed in CR is ~1.7 year. It was eported first by (Valdés-Galicia et al., 1996), analyzed by wavelet technique by (Kudela et al., 2002), found also in the outer heliosphere in Voyager data (Kato et al., 2003). Earlier, using NM data Calgary and Deep River, (Kudela et al., 1991) indicated that around ~20 month the change of the shape of PSD occurs. Recently Okhlopkov (2011) reports that length of the quasi-2 year periodicity in even and odd numbered cycles differs by ~2 months. Mendoza et al. (2006) by examining solar magnetic fluxes in the period 1971–1998 found that ~ 1.7 year is the dominant fluctuation for all the types of fluxes analyzed and that it has a strong tendency to appear

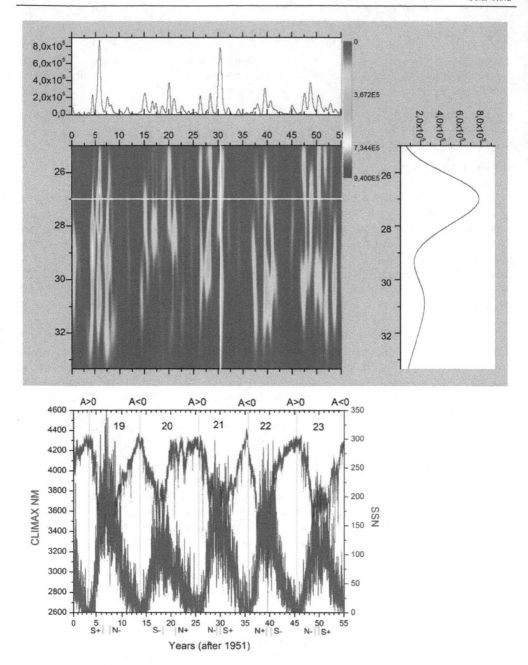

Fig. 4. The wavelet spectrum density (Morlet) of Climax NM daily means (middle panel). The upper panel is cross section of the density at 27 days over long time. Right panel is cross section over periods 25 – 33.3 d for the time of solar maxima ~ 1981. Low panel displays sunspot numbers and CR NM Climax count rates.

during the descending phase of solar activity. Quasi-periodicities of ~1.3 year (observed in solar wind) and ~1.7 years were seen neither often nor prominently in several solar activity indices (Kane, 2005). Rouillard and Lockwood (2004) relate a strong 1.68-year oscillation in GCR fluxes to a corresponding oscillation in the open solar magnetic flux and infer CR propagation paths confirming the predictions of theories in which drift is important in modulating CR flux. Charvátová (2007) indicated an interesting approach to the ~1.6 yr variation. The author calculated the solar motion due to the inner (terrestrial) planets (Mercury, Venus, Earth, Mars) for the years 1868–2030 and found that spectrum of periods shows the dominant periodicity of 1.6 years. Kane et al. (1949) reported that starting with Alfven's original suggestion, it is possible to develop a quantitative equilibrium theory for the trapping of CR in the magnetic field of the Sun, where in addition to the effect of scattering in the geomagnetic field there is also taken into account the direct absorption of the CR by five heavenly bodies Mars, Venus, the Earth, the Sun, and the Moon. This may be one of candidates for finding the link between the result (Charvátová, 2007) and ~ 1.7 year quasi-periodicity observed in CR.

At longer time scales McCracken et al. (2002) identified the presence of ~ 5 year variability in CR over epochs with low solar activity in the past. It is desirable to investigate whether a correlated 5-year signal exists in other geophysical and biological records, and if so, it could provide an additional source of data on the characteristics of the sun at times of low solar activity. El-Borie (2002) studied solar wind speed and density for quasi-periodic cyclicity and found some other long term peridicities. The 9.8, 3.8, and 1.7 – 2.2 year periods are the most significant found in the interplanetary proton flux at 190 – 440 MeV in IMP 8 data (Laurenza et al., 2009). Mavromichalaki et al (2005) reported ~2.3 years periodicity in coronal index calculated using Fe XIV 530.3 nm coronal emission line from ground-based measurements by the worldwide network of coronal stations (Rybanský, 1975). The wavelet analysis at various cut-offs for NM and for muon detector data is required to clarify whether that quasi-periodicity has a cut-off energy and how it is evolved over several tens of years. Figure 5 presents a brief summary of CR quasi-periodicities obtained in long time of measurements by NM.

Most of the quasi-periodicities identified in Figure 5 were reporteded by (Mavromichalaki et al., 2003) in the analysis of the data until 1996. For detailed analysis and for identification of contribution of various quasi-peridicities in the signal over the long time, the wavelet transform technique is suitable to be used in future, along with analysis of data from NMs at different geomagnetic cut-off positions and from muon telescopes sensitive to higher energies. The wavelet technique has been utilized also for checking the presence of other periodicities reported e.g. by (Chowdhury et al., 2010; Agarwal et al., 2011b; Zarrouk and Bennaceur, 2009) and, if used for the same intervals and applied on time series of CR as well as of solar, geomagnetic and interplanetary activity indices, may help in discriminating the links of CR to the atmospheric processes.

3. Geomagnetic effects

At low energies the trajectories of charged particles in the geomagnetic field are usually described with use of guiding center approximation (Roederer, 1970). Three adiabatic invariants connected with the three different cyclic motions, namely with the gyration, bounce and azimuthal (longitudinal) drift, are utilized. This is useful if the phases of the

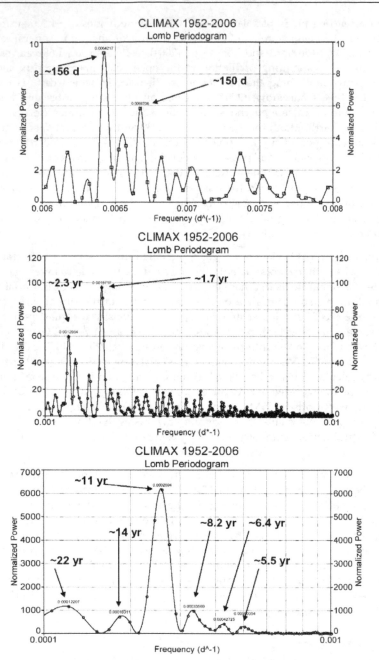

Fig. 5. Lomb-Scargle periodogram of Climax NM data at three intervals for which the filter was used. Upper panel shows the double structure with significant peaks at ~156 and ~150 days. Middle panel: in addition to ~1.7 yr variation also ~2.3 yr is statistically significant (0.99). For that level only ~11 yr variability is significant at lower plot.

motions are not of importance. Such approach is frequently used for trapped particle populations since the discovery of radiation belts. This approximation is valid if the three periodicities are very distinct. For given type of particle (proton, electron) the frequencies corresponding to the three types of motions in a dipole field depend on McIlwain's L parameter and kinetic energy (Schulz and Lanzerotti, 1974, Figure 6). The approximation becomes inapplicable when the periodicities are comparable. This is the case of higher energies in outer magnetosphere. Such particles from the point of view of magnetosphere can be assigned as CR. Detailed review of CR in the magnetospheres of planets is in (Dorman, 2009).

Trajectory description for CR particles in the Earth's magnetosphere is possible only by numerical solution of the equation of motion in given geomagnetic field model. The trajectory of particle with opposite sign of charge and velocity vector is traced starting from the point above the detector and continuing either up to the magnetosphere boundary (if the trajectory is allowed) or to the point on the ground (forbidden trajectory). For allowed trajectories the asymptotic directions are obtained. System of forbidden and allowed trajectories determines the geomagnetic cut-off rigidity. Cooke et al. (1991) summarize the definitions of the characteristics relevant for the cut-offs. Such procedure is used for long time, its history with relevant references is in the paper (Smart and Shea, 2009). Crucial for the results is the geomagnetic field model. Earlier the IGRF model was used, later the models with external current systems were introduced. Desorgher et al. (2009) discuss the geomagnetic field models used for the CR trajectory tracing. Usually for given position of NM the geomagnetic cut-off rigidity is computed for the vertical incidence of particles.

Due to geomagnetic field evolution on long time scale the geomagnetic cut-off rigidities at given point of the Earth's surface are changing (e.g. Smart and Shea, 2003; Kudela and Bobík, 2004). During the geomagnetic storms the contribution of external current systems in magnetosphere is important for the transmissivity function, asymptotic directions and geomagnetic cut-offs. Procedure that allows to determine the cut-off rigidity and asymptotic direction changes during geomagnetically active periods from measurements of magnetic field variations is presented in (Flückiger et al., 1986). Different transmissivity of CR through the Earth's magnetosphere for different empirical geomagnetic field models during strong geomagnetic storms is expected (Kudela et al., 2008). The correlations of cut-offs estimated from the global network of NMs and from trajectory tracing in one model of the field during a disturbed period, is discussed by Tyasto et al. (2011). Penetration boundary of SEP into magnetosphere is a specific tool for checking the validity of geomagnetic field models (e.g. Lazutin et al., 2011).

Recently, experiment PAMELA has shown that antiprotons produced due to nuclear interactions with the residual atmosphere are trapped in the geomagnetic field and observed at the altitude of several hundreds km (Adriani et al., 2011). Theoretical works published earlier have shown such possibility (Pugacheva et al., 2003; Gusev et al., 2008).

4. LECR, space weather and atmospheric effects

Space Weather is a relatively new discipline of science and CR play a role in this study in both aspects, namely in (a) direct one – irradiation of materials in space, in atmosphere and on the ground with various consequences for the technological systems and for the people, and (b) as one of the precursors due to changes of LECR anisotropy several hours before the onset of geomagnetic storm when CME arrives to the vicinity of Earth. LECR provide for that "remote" information about CME propagation in interplanetary space (alert for geoeffective events). There are several books and reviews on space weather effects and its possible forecasting (e.g. Song et al, 2001; Goodman, 2005; Bothmer and Daglis, 2007; Moldwin 2008; Lilensten et al., 2008), and on relations between cosmic rays, energetic particles in space and space weather (e.g. Kudela et al., 2000; 2009; Daglis (ed) 2004; Lilensten and Bornarel, 2006, Flückiger, 2007) as well as on physics behind (e.g. Scherer et al., 2005; Dorman 2010; Kallenrode, 2004; Hanslmeier, 2007; Kamide and Chian, 2007 among others). Singh et al. (2011) and Siingh (2011) reviewed CR effects on terrestrial processes such as electrical phenomena, lightning discharges cloud formation and cloud coverage, temperature variation, space weather phenomena, Earth's climate and the effects of GCRs on human health. The paper includes the new results and the authors point out many basic phenomena which require further study as well as new and long data sets.

Ions accelerated to several tens to hundreds of MeV are most important for the radiation hazard effects during solar radiation storms with electronic element failures on satellites, communication and biological consequences. Before their massive arrival, NM, if good temporal resolution and network by many stations is in real time operation, can provide useful alerts several minutes to tens minutes in advance. Kuwabara et al. (2006) report a system that detects count rate increases recorded in real time by eight NMs and triggers an alarm when GLE is detected. The GLE alert precedes the earliest indication from GOES (100 MeV or 10 MeV protons) by ~10–30 minutes. Oh el al. (2010) studied characteristics of SPE connected with GLEs.

Important point stressed by the recent papers is requirement of global detector network operating in real time with good statistical resolution is essential for space weather applications using ground based measurements. One of such systmes using neutron monitors is described by Mavromichalaki et al. (2006). At higher energies the Global Muon Detector Network is important source of the precursory information for geomagnetic storms (e.g. Rockenbach et al., 2011) and for sounding of CME geometry before its arrival to Earth (Kuwabara et al., 2004). Precursor signatures of SSC at the beginning of relatively small geomagnetic storm was also observed (Braga et al., 2011). Recently (Agarwal et al., 2011b) studied the cosmic ray, geomagnetic and interplanetary plasma/field data to understand the physical mechanism responsible for Forbush decrease and geomagnetic storm that can be used as a signature to forecast space weather and stressed the importance of change of geomagnetic cutoff rigidity.

Cosmic rays and energetic particles of lower energy interact with the material of the satellites, airplanes, atmosphere and may cause the failures. There is variety of effects with consequences on the reliability of the electronic elements. The energy deposition in materials resulting in permanent damage in silicon semiconductor devices and the single event effects due to the individual events caused by interaction of particles inside the active volume of silicon devices, along with the review of processes of electromagnetic interaction, nuclear interaction with matter is described in detail e.g. in the book (Leroy and Rancoita, 2009). The memory circuits are also partially affected by CR and its secondary products. Autran et al. (2010) review recent (2005–2010) experiments and modeling-simulation work dedicated to the evaluation of natural radiation-induced soft errors in advanced static memory (SRAM) technologies. The impact on the chip soft-error rate (SER) of both terrestrial neutrons induced by CR and alpha particle emitters, generated from traces of radioactive contaminants in CMOS process or packaging materials, has been experimentally investigated by life (i.e. real-time) testing performed at ground level on the Altitude Single-event Effect Test European Platform (ASTEP) and underground at the underground laboratory. Soft errors are caused by CR striking sensitive regions in electronic devices. Paper (Wang and Agrawal, 2010) illustrates how soft errors are a reliability concern for computer servers, and indicates a possible soft error rate (SER) reduction method that considers the CR striking angle to redesign the circuit board layout.

Miroshnichenko (2003) provides phenomenological picture of the radiation environment of Earth, summarizes observational data and theoretical findings related to main sources of energetic particles in space as well as surveys the methods of prediction of radiation risk on spacecraft. Dartnell (2011) reviews in detail the influence of ionizing radiation including CR on the emergence and persistence of life. Not only effects of ionizing radiation on organisms and the complex molecules of life are discussed, but also pointed out that ionizing radiation performs many crucial functions in the generation of habitable planetary environments and the origins of life. There are reports on the effects of short time increases of LECR on the dose within the atmosphere (airplanes, eg. Spurný et al., 2001; 2004; Felsberger et al., 2009) as well as in outer space (important for planned missions to the planets both for humans and for reliability of electronic systems); the changes of the status of the ionosphere with consequences on navigation. LECR and its measurement is important not only for monitoring radiation and its temporal and spatial variability (significant for preparing models), but its systematic measurement with good temporal resolution by many ground based devices has a potential to be one of the elements for schemes of space weather effects prediction.

Variability of CR with the aim of deducing the features useful in search of correlation between CR and atmospheric processes is described and discussed by (Bazilevskaya, 2000). Studies on relation of CR to the atmospheric processes (started probably from Svensmark and Friis-Christensen, 1997 and references therein) and references therein, recent paper reporting results of CLOUD experiment (Kirkby et al., 2011), as well as the availability of long term series of CR measurements from various NM and muon detectors until now, motivate to describe in detail LECR variability. Harrison et al. (2011) report cloud base height distributions for low cloud (<800 m) measured at the Lerwick Observatory, Shetland, UK, is varying with CR conditions. 27 day and 1.68 year periodicities characteristic of cosmic ray variations are present, weakly, in the cloud base height data of stratiform clouds, when such periodicities are present in neutron monitor CR data. Papers (Sloan and Wolfendale, 2008; Erlykin et al., 2009a,b) do not indicate that the large portion of the clouds is related to CR. No response of global cloud cover to Forbush decreases at any altitude nor latitude is reported by (Calogovic et al., 2010). Fichtner et al. (2006) point out that presence of a 22-year periodicity can not only be understood on well-known physical grounds, but must be expected if CR play a role in climate driving. The test of whether 22-year periodicities in climate indicators are present or not is a promising tool to bring the presently intensely led debate to a satisfactory conclusion. Discussion is continuing. For the purposes of checking long term variations of CR with atmospheric characteristics it is suitable to use indirectly estimated time profile of CR for the past. Usoskin et al. (2002) used the reconstructed magnetic flux as an input to a spherically symmetric quasi-steady state model of the heliosphere, and calculated the expected intensity of GCR at Earth position since 1610.

In recent decade the relations of CR to the atmospheric electricity has been studied extensively. When studying the intensity variations of secondary CR during thunderstorms (Lidvansky, 2003; Lidvansky and Khaerdinov, 2011) with the Carpet shower array of the Baksan Neutrino Observatory it was found that, in addition to regular variations correlating with the near-ground electric field, there existed considerable transient changes of the intensity. Chilingarian et al. (2010) presented the energy spectra of electrons and gamma rays from the particle avalanches produced in the thunderstorm atmosphere, reaching the Earth's surface. Paper by (Ermakov et al., 2009) shows that the main parameters of atmospheric electricity are related to CR. The mechanisms of solar forcing of the climate and long term climate change is summarized, and the role of energetic charged particles (including CR) on cloud formation and their effect on climate is discussed in (Siingh et al., 2010; 2011). Results of spectral analysis of surface atmospheric electricity data (42 years of Potential Gradient, PG at Nagycenk, Hungary) showed ~1.7 year quasi-periodicity (Harrison and Märcz, 2007). ~1.7 year periodicity in the PG data is present 1978 – 1990, but absent in 1963 – 1977. It is of interest to continue checking the occurrence of that quasi-periodicity in CR and in the data of atmospheric electricity after 1990.

Lightning is connected with the short time increases of the high energy photon flux in the atmosphere. Terrestrial Gamma–ray Flash (TGF) is a brief (<1ms) pulse of γ–rays with energies extending up to around 40 MeV, and average energy ~2 MeV (Smith et al., 2005) observed on low altitude satellites. TGFs exhibit both spatial and temporal correlations with lightning activity (e.g. Fishman et al., 1994). Study based on RHESSI data shows that on average the TGFs were found to precede the associated lightning events, with a mean delay of -0.77 ms (Collier et al., 2011). Spatial coincidence of the location of the lightning flashes with conjugate X-ray enhancements, and their simultaneity, was reported by (Bučík et al., 2006).

Cosmic ray characteristics along with the geomagnetic and solar activity are discussed also in connection with hurricanes (e.g. Kavlakov et al., 2008; Mendoza and Pazos, 2009 Perez-Peraza et al., 2008; Kane, 2006). This topic is reviewed in detail by (Mendoza, 2011).

5. Acknowledgments

We acknowledge the University of New Hampshire, "National Science Foundation Grant ATM-9912341" for Climax NM data. Author wishes to acknowledge support by the grant agency VEGA, project 2/0081/10 and by MVTS COST ES 0803 project of EU.

6. References

Abbrescia, M., S. Aiola, R. Antolini et al., Observation of the February 2011 Forbush decrease by the EEE telescopes, Eur. Phys. J. Plus, 126: 61, 2011.

Ackermann, M. et al., Measurement of separate cosmic-ray electron and positron spectra with the Fermi Large Area Telescope, arXiv:1109.0521v1, 2011.

Adriani, O. et al., PAMELA Collaboration, A statistical procedure for the identification of positrons in the PAMELA experiment, Astropart. Phys. 34, 1. arXiv:1001.3522, 2010.

Adriani, O. et al., PAMELA collaboration. Observation of an anomalous positron abundance in the cosmic Radiation, Nature, 458:607-609,2009; ArXiv:0810.4995v1, 2009.

Adriani, O. et al., The discovery of geomagnetically trapped cosmic ray antiprotons, Ap. J. Lett., 737, L29 (5pp), August 20, 2011.

Agarwal Rekha, R.K. Mishra, S.K. Pandey and P.K. Selot. 27 day variation of cosmic rays along with interplanetary parameters. Proc. 32nd ICRC, Beijing, paper 0132, 2011a.

Agarwal Rekha, Rajesh K. Mishra, M.P. Yadav and S.K. Pandey, Cosmic Rays and Space Weather Prediction, Proc. 32nd ICRC, Beijing, paper icrc0129, 2011b.

Ahluwahlia, H. S. and A.J. Dessler. Diurnal variation of cosmic radiation intensity produced by a solar wind. Planetary and Space Science, Vol. 9, p.195, 1962.

Ahluwalia, H.S. and R.C. Ygbuhay. The onset of sunspot cycle 24 and galactic cosmic ray modulation. Adv. Space Res., 48, 1, 61-64, 2011.

Ahluwalia, H.S. and S. Singh, On Higher Harmonics in Cosmic-Ray Solar Daily Variation. Proc. 13th ICRC, Denver, 2, 948, 1973.

Ahluwalia, H.S. Semidiurnal Variation of Cosmic Rays on Geomagnetically Disturbed Days. Proc. Phys. Soc. 80, 472, 1962.

Alania, M.V., Wawrzynczak, A., Energy dependence of the rigidity spectrum of Forbush decrease of galactic cosmic ray intensity, Advances in Space Research, online, 2011.

Alania, M.V., Wawrzynczak, A., Forbush decrease of the Galactic Cosmic Ray Intensity: Experimental Study and Theoretical Modeling, Astrophys. And Space Sci. Transactions, 4, 59-63, 2008.

Andriopoulou, M., H. Mavromichalaki, C. Plainaki, A. Belov and E. Eroshenko. Intense Ground-Level Enhancements of Solar Cosmic Rays During the Last Solar Cycles, Solar Phys., 269, 155-168, 2011.

Augusto, C.R.A., C. E. Navia, H. Shigueoka, and K. H. Tsui. Muon excess at sea level from solar flares in association with the Fermi GBM spacecraft detector. Phys. Rev. D 84, 042002, 2011.

Autran, J.L., D. Munteanu, P. Roche, G. Gasiot, S. Martinie, S. Uznanski, S. Sauze, S. Semikh, E. Yakushev, S. Rozov, P. Loaiza, G. Warot, M. Zampaolo. Soft-errors induced by terrestrial neutrons and natural alpha-particle emitters in advanced memory circuits at ground level. Microelectronics Reliability 50, 1822–1831, 2010.

Badruddin. Solar modulation during unusual minimum of solar cycle 23: Comparison with past three solar minima, Proc. 32nd ICRC, Beijing, paper icrc0116, 2011.

Balogh, A. and G. Erdös. The heliospheric magnetic field. Space Sci. Rev., published online, DOI 10.1007/s11214-011-9835-3, 2011.

Barnard, L. and M. Lockwood. A survey of gradual solar energetic particle events. JGR, 116, A05103, 2011.

Bazilevskaya, G.A. Observations of Variability in Cosmic Rays, Space Science Rev., 94, 25-38, 2000.

Bazilevskaya, G.A. , Krainev, M.B., Makhmutov, V.S., Svirzhevskaya, A.K., Svirzhevsky, N.S., Stozhkov, Y.L. , Features of cosmic ray variation at the phase of the minimum between the 23rd and 24th solar cycles, Bull. Russian Acad. Sci: Physics , 75, 6, 779-781, 2011.

Belov, A. Large Scale Modulation: View From the Earth. Space Sci. Rev., 93,1/2, 79-105, 2000.

Bertou, X. Background radiation measurement with water Cherenkov detectors, Nucl. Instruments and Methods in Physics research, section A, 639, 1, 73-76, 2011.

Bieber, J.W.; Eroshenko, E.; Evenson, P.; Flückiger, E.O.; Kallenbach, R. , Editors. Cosmic Rays and Earth. Proceedings of an ISSI Workshop 21-26 March 1999, Bern, Switzerland, Series: Space Sciences Series of ISSI, Vol. 10, 2001.

Biermann, L., O. Haxel, A. Schluter. Neutrale Ultrastrahlung von der Sonne. Z. Naturforsch. 6a, 47, 1951.

Bothmer, V. and I.A. Daglis. Space Weather – Physics and Effects. Springer, pp. 437, 2007.

Braga, C.R., A. Dal Lago, M. Rockenbach, N.J. Shuch, L.R. Vieira, K. Munakata, C. Kato, T. Kuwabara, P.A. Evenson, J. W. Bieber, M. Tokumaru, M.L. Duldig, J.E. Humble, I.S. Sabbah, H.K. Al Jassar, M.M. Sharma. Precursor signatures of the storm sudden commencement in 2008. Proc. 32nd ICRC, Beijing, paper icrc0717.

Braun, I., J. Engler, J.R. Horandel and J. Milke. Forbush decreases and solar events seen in the 10–20 GeV energy range by the Karlsruhe Muon Telescope, Advances in Space Research 43, 480–488, 2009.

Brunberg, E. A. and A. Dattner. On the Interpretation of the Diurnal Variation of Cosmic Rays. Tellus, 6, 1, 73-83, 1954.

Bučík, R., K. Kudela and S.N. Kuznetsov. Satellite observations of lightning-induced hard X-ray flux enhancements in the conjugate region, Ann. Geophys., 24, 1969–1976, 2006.

Burlaga, L.F. and N.F. Ness, Magnetic field strength distributions and spectra in the heliosphere and their significance for cosmic ray modulation: Voyager 1, 1980-1994, JGR, 103, A12, 29,719-29,732, 1998.

Burlaga, L.F., F.B. McDonald, M.L. Goldstein and A.J. Lazarus, Cosmic ray modulation and turbulent interaction regions near 11 AU, J. Geophys. Res., 90, A12, 12,027-12,039, 1985.

Calogovic, J., C. Albert, F. Arnold, J. Beer, L. Desorgher, and E. O. Flückiger, Sudden cosmic ray decreases: No change of global cloud cover, Geophys. Res. Lett., 37, L03802, doi:10.1029/2009GL041327, 2010.

Cane, H.V., G. Wibberenz, I.G. Richardson and T.T. von Rosenvinge, Cosmic ray modulation and the solar magnetic field, Geophys. Res. Lett., 26, 5, 565-568, 1999.

Cane, H.V., I.G. Richardson and T.T. von Rosenvinge, Interplanetary magnetic field periodicity of ~153 days, GRL, 25, 4437-4440, 1998.

Collier, A.B., T. Gjesteland and N. Østgaard, Assessing the power law distribution of TGFs. J. Geophys. Res., 116, A10320, doi:10.1029/2011JA016612, 2011.

Cooke, D.J., J.E. Humble, M.A. Shea, D.F. Smart, N. Lund, I.L. Rasmussen, B. Byrnak, P. Goret and N. Petrou. On cosmic-ray cut-off terminology. Il Nuovo Cimento 14, 213-234, 1991.

Daglis, I.A. (editor), Effects of Space Weather on Technology Infrastructure, Proc. of the NATO ARW on Effects of Space Weather on Technology Infrastructure, Rhodes, Greece, 2003, Kluwer Academic Publishers, 2004.

Dartnell, L.R. Ionizing Radiation and Life. Astrobiology, 11, 6, 551-582, 2011.

Debrunner, H., E.O. Flückiger, E.L. Chupp & D.J. Forrest. The solar cosmic ray neutron event on June 3, 1982. Proc. 18th ICRC,Bangalore, India, 4, 75–78, 1983.

DeMendonca, R.R.S., J.-P.Raulin, F.C.P.Bertoni, E.Echer, V.S.Makhmutov, G.Fernandez, Long-term and transient time variation of cosmic ray fluxes detected in Argentina by CARPET cosmic ray detector, J. Atmos. Solar Terr. Physics, 73, 1410-1416, 2011.

Desorgher, L., Kudela, K., Flückiger, E. O., Bütikofer, R., Storini, M., Kalegaev, V., Comparison of Earth's magnetospheric magnetic field models in the context of cosmic ray physics, Acta Geophysica, 57, 1, 75-87, 2009.

Dorman, L.I. Cosmic ray variations and space weather, Physics – Uspekhi, 53 (5), 496-503, 2010b.

Dorman, L.I. Variations of galactic cosmic rays, Moscow, MGU Publ. House, pp. 214, in Russian, 1975.

Dorman, Lev. Cosmic Rays in Magnetospheres of the Earth and other Planets, Astrophysics and Space Science Library, 358, Springer, pp. 770, 2009.

Dorman, Lev. Cosmic Rays in the Earth's Atmosphere and Underground. Astrophysics and Space Science Library, pp. 855, Kluwer, 2004.

Dorman, Lev. Cosmic Rays: Variations and Space Explorations. Elsevier Science Publishing Co Inc.,U.S. pp. 691, 1974.

Dorman, Lev. Solar Neutrons and Related Phenomena, Astrophysics and Space Science Library, 365, Springer, pp. 873, 2010a.

Duldig, M.L., Australian Cosmic Ray Modulation Research. Publ. Astron. Soc. Austr., 18, 12-40, 2001.

Dumbovic, M., B. Vrsnak, J. Calogovic, and M. Karlica, Cosmic ray modulation by solar wind disturbances, A&A 531, A91, 2011.

Dunzlaff, P., Heber, B., Kopp, A., Rother, O., Müller-Mellin, R., Klassen, A., Gómez-Herrero, R., Wimmer-Schweingruber, R. Observations of recurrent cosmic ray decreases during solar cycles 22 and 23, Ann. Geophys., 26, 3127-3138, 2008.

Dvornikov, V.M. and V. E. Sdobnov, Correction of Data from the Neutron Monitor Worldwide Network, Geomagnetism and Aeronomy, Vol. 48, No. 3, pp. 314–318, 2008.

Efimov, Yu. E.; Kocharov, G. E.; Kudela, K. On the solar neutrons observation on high mountain neutron monitor. Proc. 18th ICRC, Bangalore, India, 10, 276 – 278, 1983.

El-Borie, M.A. and S.S. Al-Thoyaib, Power Spectrum of Cosmic-ray Fluctuations During Consecutive Solar Minimum and Maximum Periods. Solar Phys., 209, 397–407, 2002.

El-Borie, M.A. On Long-Term Periodicities In The Solar-Wind Ion Density and Speed Measurements During The Period 1973-2000. Solar Phys., 208, 345–358, 2002.

Erlykin, A.D.; Gyalai, G.; Kudela, K.; Sloan, T.; Wolfendale, A.W., On the correlation between cosmic ray intensity and cloud cover, Journal of Atmospheric and Solar-Terrestrial Physics, Volume 71, Issue 17-18, 1794-1806, 2009a.

Erlykin, A.D., G Gyalai, K Kudela, T Sloan, A W Wolfendale. Some aspects of ionization and the cloud cover, cosmic ray correlation problem, Journal of Atmospheric and Solar-Terrestrial Physics, 71, 8-9, 823-829, 2009b.

Ermakov, V.I., V. P. Okhlopkov and Yu. I. Stozhkov, Influence of cosmic rays and cosmic dust on the atmosphere and Earth's climate, Bull. Rus. Acad. Sci., Physics, 73, 3, 416-418, DOI: 10.3103/S1062873809030411, 2009.

Evenson, P.; Meyer, P.; Pyle, K. R., Protons from the decay of solar flare neutrons, Astrophysical Journal, Part 1, 274, 875-882, 1983.

Felsberger, E., K. O'Brien, P. Kindl. : Iason-free: Theory and experimental comparisons. Radiation Protection Dosimetry, Vol. 136, Issue 4, 16 July 2009, Article number ncp128, 267-273, 2009.

Fichtner, H., K. Scherer and B. Heber. A criterion to discriminate between solar and cosmic ray forcing of the terrestrial climate. Atmos. Chem. Phys. Discuss., 6, 10811–10836, 2006.

Filisetti, O. and V. Mussino, Periodicity of about 13 days in the cosmic-ray intensity in the solar cycles no. 18, 19 and 20. Rev. Bras. Fis., Vol. 12, No. 4, p 599 – 610, 12/1982.

Fishman, G. J., et al. (1994), Discovery of intense gamma–ray flashes of atmospheric origin, Science, 264(5163), 1313–1316, doi:10.1126/science.264.5163.1313, 1994.

Fisk, L. A.; M.A. Forman and W.I. Axford. Solar modulation of galactic cosmic rays. 3. Implication of the Compton-Getting coefficient. JGR, 78, 995-1006, 1973.

Flückiger, E.O. Cosmic Rays and Space Weather, presentation at http://www.slidefinder.net/c/cosmic_rays_and/space_weather/1510213, 2007.

Flückiger, E. O., D.F. Smart, M.A. Shea. A procedure for estimating the changes in cosmic ray cutoff rigidities and asymptotic directions at low and middle latitudes during periods of enhanced geomagnetic activity. Journal of Geophysical Research 91, 7925-7930, 1986.

Gabriel, S., R. Evans, and J. Feynman, Periodicities in the occurrence rate of solar proton events, Sol. Phys., 128, 415-422, 1990.

Gaisser, T.K. Cosmic Rays and Particle Physics. Cambridge University Press, pp. 279, 1990.

Garcia-Munoz, M., G.M. Mason and J.A. Simpson. A New Test for Solar Modulation Theory: the 1972 May-July Low-Energy Galactic Cosmic-Ray Proton and Helium Spectra. ApJ., 182, L81, 1973.

Gil, A., Iskra, K., Modzelewska, R., Alania, M. V., On the 27-day variations of the galactic cosmic ray anisotropy and intensity for different periods of solar magnetic cycle, Adv. Space Res., 35, 687-690, 2005.

Gleeson, L. J. and Axford, W.I. Solar modulation of galactic cosmic rays, The Ap. J., 154, 1011-1026, 1968.

Goodman, J.M. Space Weather & Telecommunications, Kluwer International Series in Engineering and Computer Science, Springer,pp.382, 2005.

Gopalswamy, N. A Global Picture of CMEs in the Inner Heliosphere. In The Sun and the heliosphere as an integrated system. Editors G. Poletto and S.T. Suess, Kluwer, 201-252, 2004.

Gopalswamy, N., H. Xie, S. Yashiro and I. Usoskin. Ground level enhancement events of solar cycle 23, Indian Journal of Radio & Space Physics, 39, 240-248, 2010.

Grieder, P.K.F. Cosmic Rays at Earth. Researcher's Reference Manual and Data Book, pp. 1093. Elsevier, 2001.

Grieder, P.K.F. Extensive Air Showers, vol. 1 and 2, pp. 1113, Springer, 2010.

Grimani, C., H.M. Araujo, M. Fabi, I. Mateos, D.N.A. Shaul, T.J. Sumner and P. Wass. Galactic cosmic-ray energy spectra and expected solar events at the time of future space missions. Classical and Quantum Gravity, 28, 9, Art. No. 094005, 2011.

Grupen, K. Astroparticle Physics. Springer Berlin Heidelberg New York, pp. 441, 2005.

Gusev, A.A., G. I. Pugacheva, V. Pankov, J. Bickford, W. N. Spjeldvik, U. B. Jayanthi and I. M. Martin, Antiparticle Content in the Magnetosphere, Advances in Space Research, Volume 42, Issue 9, 3 November 2008, Pages 1550-1555, 2008.

Hanslmeier, A, The Sun and Space Weather, Astrophysics and Space Physics Library, 347, Springer, pp. 315, 2007.

Harrison, R.G. and F. Märcz, Heliospheric timescale identified in surface atmospheric electricity, GRL, 34, L23816, 2007GL031714, 2007.

Harrison, R.G., M.H. P. Ambaum and M. Hapgood, Cloud base height and cosmic rays, Proc. R. Soc. A doi:10.1098/rspa.2011.0040, 2011.

Hatton, C.J. The Neutron Monitor, in J., G. Wilson and S.A. Wouthuysen (eds.), Progress in Elementary Particle and Cosmic-ray Physics, vol. 10, chapter 1, North Holland Publishing Co., Amsterdam, 1971.

Hayakawa, S. Cosmic ray physics: nuclear and astrophysical aspects, Wiley-Interscience, U. California, pp. 774, 1969.

Hess, V.F. Über beobachtungen der durchdringenden Strahlung bei sieben Freiballonfahrten, Phys. Ztschr., 13, 1084-1091, 1912.

Hill, M.E., D.C. Hamilton and S.M. Krimigis, Radial and Latitudinal Intensity Gradients of Anomalous Cosmic Rays During the Solar Cycle 22 Recovery Phase, JGR, 106, A5,8315, 2001.

Hillas. A.M. Cosmic Rays. Pergamon Press, pp. 297, 1972.

Hudson, H.S. Global properties of solar flares. Space Sci. Rev., 158, 5-41, 2011.

Charvátová, I., The prominent 1.6-year periodicity in solar motion due to the inner planets, Ann. Geophys., 25, 1227–1232, 2007.

Chauhan, M.L., Manjula Jain and S. K. Shrivastava. Space weather application of forbush decrease events. Proc. 32nd ICRC Beijing, paper icrc0155, 2011.

Chilingarian, A., A. Daryan, K. Arakelyan, A. Hovhannisyan, B. Mailyan, L. Melkumyan, G. Hovsepyan, S. Chilingaryan, A. Reymers, and L. Vanyan, Ground-based observations of thunderstorm-correlated fluxes of high-energy electrons, gamma rays, and neutrons. Phys. Rev. D 82, 043009, 2010.

Chowdhury, P., B.N. Dwivedi and P.C. Ray. Solar modulation of galactic cosmic rays during 19–23 solar cycles, New Astronomy, 16, 430-438, 2011.

Chowdhury, Partha; Khan, Manoranjan; Ray, P. C., Evaluation of the intermediate-term periodicities in solar and cosmic ray activities during cycle 23, Astrophys. Space Sci., 326, 191-201, 2010.

Chupp, E. L., D.J. Forrest, P.R. Higbie, A.N. Suri, C.Tsai and P.P. Dunphy. Solar Gamma Ray Lines observed during the Solar Activity of August 2 to August 11, 1972. Nature, 241, 5388, 333-335, 1973.

Jokipii, J. R., Cosmic-ray propagation, 2, Diffusion in the interplanetary magnetic field, Astrophys. J., 149, 405, 1967.

Jokipii, J. R., Cosmic-ray propagation, 1, Charged particles in a random magnetic field, Astrophys. J., 146, 480, 1966.

Jokipii, J.R. and A.J. Owens, Cross correlation between cosmic-ray fluctuations and interplanetary magnetic-field fluctuations, GRL, 1,329, 1974b.

Jokipii, J.R. and A.J. Owens, Cosmic Ray Scintillations, 4. The Effects of Non-Field-Aligned Diffusion. JGR, 81, 13, 2094-2096, 1974a.

Jokipii, J.R. Cosmic Rays. In Auroras, Magnetic Storms, Solar Flares, Cosmic Rays. Ed. Suess, S.T. & Tsurutani, B., AGU, 123-13, 1998.

Kallenrode, May-Britt. Space Physics: An Introduction to Plasmas and Particles in the Heliosphere and Magnetospheres, Springer-Verlag Berlin, Heidelberg, pp. 482, 2004.

Kamide, Y. and A. Chian, editors. Handbook of the Solar-Terrestrial Environment. Springer, pp. 539, 2007.

Kane, E.O., J.B. Shanley and J.A. Wheeler,Influence on the Cosmic-Ray Spectrum of Five Heavenly Bodies, *Rev. Mod. Phys.*, 21, 1, 51-71, 1949.

Kane, R.P. Severe geomagnetic storms and Forbush decreases: interplanetary relationships reexamined, Ann. Geophys., 28, 479–489, 2010.

Kane, R.P. Spectral characteristics of Atlantic seasonal storm frequency. J. of India Meteor. Dept. (MAUSAM), 57, 597-608, 2006.

Kane, R.P., Short-Term Periodicities in Solar Indices, *Sol. Phys.*, 227, 155-175, 2005.

Kato, C.; Munakata, K.; Yasue, S.; Inoue, K.; McDonald, F. B., A ~1.7-year quasi-periodicity in cosmic ray intensity variation observed in the outer heliosphere. *JGR*, 108, A10, 1367, 2003.

Kavlakov, S., J. Perez-Peraza and J.B. Elsner, A statistical link between tropical cyclone intesification and major geomagnetic disturbances, Geofisica Internacional, 47, 207-213, 2008.

Kecskeméty, K., Yu. I. Logachev, M. A. Zeldovich and J. Kota, Modulation of the galactic low energy proton spectrum in the inner heliosphere, The Astrophysical Journal, 738:173 (10pp), 2011 September 10.

Kirkby, J. et al., Role of sulphuric acid, ammonia and galactic cosmic rays in atmospheric aerosol nucleation, Nature 476, 429–433 (25 August 2011).

Kóta, J. and J.R. Jokipii. Effects of drift on the transport of cosmic rays VI. A three dimensional model including diffusion. Astrophys. J. , 265, 573, 1983.

Krymsky, G.F., V.P. Mamrukova, P.A. Krivoshapkin, S.K. Gerasimova, S.A. Starodubtsev. Recurrent variations in the high-energy cosmic ray intensity. Proc. 30th ICRC, Mexico, v. 1, 381-384, 2008.

Kudela, K., A.G. Ananth and D. Venkatesan. The low-frequency spectral behavior of cosmic ray intensity. JGR, 96, 15,871-15,875, 1991.

Kudela, K. and P. Bobík. Long-Term variations of geomagnetic rigidity cutoffs, Sol. Phys., 224, 1-2, 423-431, 2004.

Kudela, K. and R. Brenkus. Cosmic ray decreases and geomagnetic activity: list of events 1982–2002. J. Atmos. Solar Terr. Phys., 66, 1121–1126, 2004.

Kudela, K. and R. Langer. Ground Level Events Recorded at Lomnicky Stit Neutron Monitor. Proc. 30th ICRC, Mérida, Mexico. 1, SH.1.8, 205-208, 2008.

Kudela, K., Cosmic Rays and Space Waether: Direct and Indirect Relations, Proc. 30th ICRC, Mexico, ed. R. Caballero, J.C. D'Olivo, G. Medina-Tanco and J.F. Valdés-Galicia, UNAM, 195-208, 2009.

Kudela, K., Venkatesan, D., Flückiger, E.O., Martin, I.M., Slivka, M. and H. Graumann, Cosmic Ray Variations: Periodicities at T<24 hours, Proc. 24th ICRC, Rome, vol. 4, p. 928-931, 1995.

Kudela, K., Mavromichalaki, H., Papaioannou, A., Gerontidou, M., On Mid-Term Periodicities in Cosmic Rays, Sol. Phys., 266, 173–180, 2010.

Kudela, K., Storini, M., Hofer, M.Y., Belov, A. Cosmic rays in relation to space weather. Space Sci Rev 93(1–2):153–174, 2000.

Kudela, K.; Rybák, J.; Antalová, A.; Storini, M., Time Evolution of low-Frequency Periodicities in Cosmic ray Intensity, *Sol. Phys.*, 205, 165 – 175, 2002.

Kudela, K. et al., On quasi-periodic variations in cosmic rays, submitted for Proc. 13th ICATPP, Como, Italy, 2011.

Kurt, Victoria, B. Yushkov, K. Kudela, and V. Galkin. High-Energy Gamma Radiation of Solar Flares as an Indicator of Acceleration of Energetic Protons. Cosmic Research, 48, 1, 70-79, 2010.

Kurt, Victoria, B. Yushkov, A. Belov, I. Chertok and V. Grechnev. A Relation between Solar Flare Manifestations and the GLE Onset, Proc. 32nd ICRC, Beijing, paper icrc0441, 2011.

Kuwabara, T., et al. Geometry of an interplanetary CME on October 29, 2003 deduced from cosmic rays, Geophys. Res. Lett., 31, L19803, doi:10.1029/2004GL020803, 2004.

Kuwabara, T., J. W. Bieber, J. Clem, P. Evenson, and R. Pyle . Development of a ground level enhancement alarm system based upon neutron monitors , Space Weather, 4, S10001, SW000223, 2006.

Kuznetsov, S.N., V.G. Kurt, B.Y. Yushkov, K. Kudela and V.I. Galkin. Gamma-Ray and High-Energy -Neutron Measurements on CORONAS-F during the Solar Flare of 28 October 2003. Sol. Phys., 268, 1, 175-193, 2011.

Kuznetsov, S.N., V.G. Kurt, I.N. Myagkova, B. Y. Yushkov and K. Kudela, Gamma-ray emission and neutrons from solar flares recorded by the SONG instrument in 2001–2004, Solar System Res., 40, 2, 104-110, 2006.

Laurenza, M.; Storini, M.; Giangravè, S.; Moreno, G., Search for periodicities in the IMP 8 Charged Particle Measurement Experiment proton fluxes for the energy bands 0.50-0.96 MeV and 190-440 MeV, JGR, 114, A01103, 2009.

Lazutin, L.L., E.A.Muraveva, K.Kudela and M.Slivka : Verification of Magnetic Field Models Based on Measurements of Solar Cosmic Rays Protons in the Magnetosphere, Geomagnetism and Aeronomy, 51, 2, 198-209, 2011.

Leroy Claude and Pier-Giorgio Rancoita. Principles of Radiation Interacction in Matter and Detection, 2nd Edition. World Scientific, pp. 930, 2009.

Lidvansky, A.S. and N.S. Khaerdinov, Cosmic Rays in Thunderstorm atmosphere: variations of different components and accompanying effects, in press, Proc. 13th ICATPP, Como, Italy, 2011.

Lidvansky, A.S. The Effect of the Electric Field of the Atmosphere on Cosmic Rays, J. Phys. G: Nucl. Part. Phys., vol. 29, pp. 925-937, 2003.

Lilensten, J. and J. Bornarel, Space Weather, Environment and Societies, Springer, Dordrecht, The Netherlands, pp. 241, 2006.

Lilensten, J., A. Belahaki, M. Messerotti, R. Vainio, J. Watermann and Stefaan Poedts, editors. Development the scientific basis for monitoring, modelling and predicting Space Waetaher. COST Office, Brussels, pp. 359, 2008.

Lin, R. P., Solar particle acceleration and propagation, Rev. Geophys., 25, 676, 1987.

Longair, M. High Energy Astrophysics. Cambridge University Press, 1981.

Maghrabi, A.H., Al Harbi, H., Al-Mostafa, Z.A., Kordi, M.N., Al-Shehri, S.M., The KACST muon detector and its application to cosmic-ray variations studies, Advances in Space Research, doi: 10.1016/j.asr.2011.10.011, 2011.

Mavromichalaki, H., G. Souvatzoglou, C. Sarlanis, G. Mariatos, C. Plainaki, M. Gerontidou, A. Belov, E. Eroshenko, V. Yanke, Space weather prediction by cosmic rays, Advances in Space Research, 37, 1141–1147, 2006.

Mavromichalaki, I.I., Preka-Papadema, P., Petropoulos, B., Tsagouri, I., Georgakopoulos, S., and Polygiannakis, J. Low- and high-frequency spectral behavior of cosmic-ray intensity for the period 1953-1996. Ann. Geophys., 21, 1681-1689, 2003.

Mavromichalaki, H.; Petropoulos, B.; Plainaki, C.; Dionatos, O.; Zouganelis, I., Coronal index as a solar activity index applied to space weather, Adv. Space Res. ,35, 410–415, 2005.

McCracken, K.G.,Beer, J. and McDonald, F.B., A five-year variability in the modulation of the galactic cosmic radiation over epochs of low solar activity, GRL, 29, NO. 24, 2161, 2002.

Mendoza, B. and M.A. Pazos. A 22-yrs hurricane cycle and its relation to geomagnetic activity, J. Atm. Solar-Terr. Phys., 71, 17-18, 2047-2054, 2009.

Mendoza, B. The effects of space weather on hurricane activity. INTECHopen, ed. A. Lupo, April 2011 (http://www.intechopen.com/articles/show/title/the-effects-of-space-weather-on-hurricane-activity).

Mendoza, B.V., V. M. Velasco and J. F. Valdés-Galicia, Mid-Term Periodicities in the Solar Magnetic Flux, Sol. Phys., 233, Issue 2, pp.319-330, 2006.

Miroshnichenko, L.I. Radiation Hazard in Space. Astrophys. And Space Science Library 207, Kluwer, Dordrecht, pp. 238, 2003.

Miroshnichenko, L.I. Solar Cosmic Rays. Astrophysics and Space Physics Library, 260, pp. 480, Kluwer, 2001.

Mishra, B.K., P. J. Shrivastava and R.K. Tiwari. A Study of the Role of the Coronal Mass Ejections in Cosmic Ray Modulation, J. Pure Appl. & Ind. Phys. Vol.1 (4), 222-226, 2011.

Moldwin, M. An Introduction to Space Weather. Cambridge U. Press, pp. 134, 2008.

Moraal, H. and K.G. McCracken. The Time Structure of Ground Level Enhancements in Solar Cycle 23, Space Sci Rev., DOI 10.1007/s11214-011-9742-7, online 2011.

Moraal, H. and Stoker, P. H.: Long-term neutron monitor observations and the 2009 cosmic ray maximum, Geophys. Res., 115, A12109, 2010.

Mori, S., S. Yasue and M. Ichinose, The Daily-Variation Third Harmonic of the Cosmic Radiation., paper MOD-37, Proc. 12th ICRC, Hobart, 2, 666, 1971.

Mufti, S., M.A.Darzi, P.M.Ishtiaq, T.A.Mir and G.N.Shah, Enhanced diurnal variation and Forbush decreases recorded with Lead-Free Gulmarg Neutron Monitor during the solar active period of late October 1989, Planet. Space Sci., 59, 394-401, 2011.

Mulligan, T., J.B. Blake, D. Shaul, J.J. Quenby, R.A. Leske, R.A. Mewaldt and M. Galamertz, Short-period variability in the galactic cosmic ray intensity: High statistical resolution observations and interpretation around the time of a Forbush decrease in August 2006, JGR, 114, A07105, 2009.

Mustajab, F. and Badruddin, Geoeffectiveness of the interplanetary manifestations of coronal mass ejections and solar-wind stream–stream interactions, Astrophys Space Sci (2011) 331: 91–104, 2011.

Nagashima, K., K. Fujimoto, S. Sakakibara, I. Morishita, R. Tatsuoka. Local-time-dependent pre-IMF-shock decrease and post-shock increase of cosmic rays, produced respectively by their IMF-collimated outward and inward flows across the shock responsible for forbush decrease, Planetary and Space Science, 40, 8, 1109-113, 1992.

Newkirk, G., Jr.; Hundhausen, A. J.; Pizzo, V., Solar cycle modulation of galactic cosmic rays - Speculation on the role of coronal transients, JGR, 86, 5387-5396, 1981.

Nicolson, P. and V. Sarabhai, The Semi-Diurnal Variation in Cosmic Ray Intensity, Proc. Phys. Soc. 60, 509, 1948.

Oh, S.Y. and Y. Yi. Statistical reality of globally nonsimultaneous Forbush decrease events. JGR, 114, A11102, 2009.

Oh, S.I., Y. Yi, J.W. Bieber, P. Evenson and Y.K. Kim, Characteristics of solar proton events associated with ground level enhancements. JGR, 115, A10107, 2010.

Okhlopkov, V.P., Distinctive properties of the frequency spectra of cosmic ray variations and parameters of solar activity and the interplanetary medium in solar cycles 20-23, Moscow U. Phys. Bull, 66, 1, 99–103, 2011.

Okike, O. and A.B. Collier. A multivariate study of Forbush decrease simultaneity. Journal of Atmospheric and Solar-Terrestrial Physics, 73, 7-8, 796-804, 2011.

Palmer, I.D. Transport coefficients of low energy cosmic rays in interplanetary space, Rev. Geophysics, 20, 2, 335-351, 1982.

Pap, J., W.K. Tobiska and S.D. Bouwer, Periodicities of solar irradiance and solar activity indices, Sol. Phys., 129, 165-189, 1990.

Parker, E.N. 1965. The passage of energetic charged particles through interplanetary space. Planet. Space Sci., 13, 1, 9-49, 1965.

Pérez-Peraza, J., Velasco, V. and S. Kavlakov. Wavelet coherence analysis of atlantic hurricanes and cosmic rays. Geofísica Internacional, 47, 231-244, 2008.

Pintér, T., M. Rybanský, K. Kudela, I. Dorotovič, Peculiarities in evolutions of cosmic radiation level after sudden decreases, in press, Sun and Geosphere, 2011.

Potgieter, M. S. & Le Roux, J. A, The Long-Term Heliospheric Modulation of Galactic Cosmic Rays according to a Time-dependent Drift Model with Merged Interaction Regions, Astrophysical Journal v.423, p.817-827, 1994.

Potgieter, M.S. and H. Moraal, A drift model for the modulation of galactic cosmic rays. Astrophys. J., 294, 425–440, 1985.

Potgieter, M.S. Cosmic Rays in the Inner Heliosphere: Insights from Observations, Theory and Models, Space Sci Rev., DOI 10.1007/s11214-011-9750-7, online, 2011.

Prasad Subramanian. Forbush decreases and Space weather. Available at http://www.iiap.res.in/ihy/school/prasad_lecture.pdf, 2009.

Pugacheva, G. I., A. A. Gusev, U. B. Jayanthi, N. G. Schuch, W. N. Spjeldvik, and K. T. Choque, Antiprotons Confined in the Earth's Inner Magnetosphere, Astroparticle Physics, 20, p.257-265, 2003.

Quenby, J.J., T. Mulligan, J.B. Blake, J.E. Mazur and D. Shaul. Local and nonlocal geometry of interplanetary coronal mass ejections: Galactic cosmic ray (GCR) short-period variations and magnetic field modeling, JGR, 113, A10102, 2008.

Ramaty, R., B. Kozlovsky and R.E. Lingenfelter. Nuclear gamma-rays from energetic particle interactions, Astrophys. J. Suppl. Ser., 40, 487-526, 1979.

Reames, D.V. Particle acceleration at the Sun and in the heliosphere. Space Science Reviews, 90, 3/4, 413-491, 1999.

Rieger, E., Kanbach, G., Reppin, C., Share, G. H., Forrest, D. J., Chupp, E. L., A 154-day periodicity in the occurrence of hard solar flares?, Nature, 312, 623-625, 1984.

Richardson, I.G. and H. V. Cane. Galactic Cosmic Ray Intensity Response to Interplanetary Coronal Mass Ejections/Magnetic Clouds in 1995-2009. Sol. Phys., 270, 2, 609-627, 2011.

Richardson, I.G. Energetic Particles and Corotating Interaction Regions in the Solar Wind. Space Sci. Rev., 111, 267-376, 2004.

Rockenbach, M., A. Dal Lago, W. D. Gonzalez, K. Munakata, C. Kato, T. Kuwabara, J. Bieber, N. J. Schuch, M. L. Duldig, J. E. Humble, H. K. Al Jassar, M. M. Sharma, and I. Sabbah, Geomagnetic storm's precursors observed from 2001 to 2007 with the Global Muon Detector Network (GMDN), GRL, 38, L16108, doi:10.1029/2011GL048556, 2011.

Roederer, J.G. Dynamics of Geomagnetically Trapped Radiation. Springer, pp. 166, 1970.

Rossi, B. and S. Olbert. Introduction to the Physics of Space. McGraw-Hill Book Co., pp. 454, 1970.

Rouillard, A. and M.A. Lockwood, Oscillations in the open solar magnetic flux with a period of 1.68 years: imprint on galactic cosmic rays and implications for heliospheric shielding, Ann. Geophys., 22, 4381-4395, 2004.

Rozelot, J.-P. Solar and Heliospheric Origins of Space Weather Phenomena. Lecture Notes in Physics 699, Springer, pp. 166, 2006.

Rybanský, M., Coronal index of solar activity. I - Line 5303 A, year 1971. II - Line 5303 A, years 1972 and 1973, Bull. Astron. Inst. Czech. 26, 367-377, 1975.

Sabbah, I. and K. Kudela, Third harmonic of the 27 day periodicity of galactic cosmic rays: Coupling with interplanetary parameter, JGR, 116, A04103, 2011.

Sabbah, I. and M.L. Duldig, Solar Polarity Dependence of Cosmic Ray Power Spectra Observed with Mawson Underground Muon Telescopes. Solar Phys., 243, 231-235, 2007.

Sdobnov, V.E., Analysis of Forbush effect in May 2005 by the method of spectrographic global survey, (in Russian), Izv. RAN, ser. Phys., 75, 6, 872-874, 2011.

Serpico, P.D.. Astrophysical models for the origin of the positron "excess". Astroparticle Physics, in press, arXiv:1108.4827v1, 2011.

Scherer, K., H. Fichtner, B. Heber, U. Mall (eds.), Space Weather: The Physics Behind a Slogan, Lecture Notes in Physics, Springer Berlin and Heidelberg, 2005, pp. 297, 2005.

Schulz, M. and L.J. Lanzerotti. Particle Diffusion in the Radiation Belts. Springer, pp. 215, 1974.

Siingh, D., Singh, R.P. Singh, A.K. , Kulkarni, M.N. , Gautam, A.S. , Singh, A.K,. Solar Activity, Lightning and Climate, Surveys in Geophysics, 32, 6, 659-703, 2011.

Siingh, Devendraa; Singh, R. P., The role of cosmic rays in the Earth's atmospheric processes, Pramana, vol. 74, issue 1, pp. 153-168, 2010.

Siluszyk, M., A. Wawrzynczak and M.V. Alania. A model of the long period galactic cosmic ray variations. Journal of Atmospheric and Solar-Terrestrial Physics, Volume 73, Issue 13, 1923-1929, 2011.

Simpson, J.A. Cosmic Radiation Neutron Intensity Monitor, Annals of the Int. Geophysical Year IV, Part VII, Pergamon Press, London, p. 351, 1958.

Singer, S.F. Cosmic Rays and the Sun's Magnetic Field: Diurnal Variation of Cosmic Rays and the Sun's Magnetic Field. Nature, 170, 4315, 63-64, 1952.

Singh, A.K., Devendraa Siingh, Singh, R.P. Impact of galactic cosmic rays on Earth's atmosphere and human health, Atmos. Environment, 3806-3818, 2011.

Sloan, T. and A.W. Wolfendale, Testing the proposed causal link between cosmic rays and cloud cover, Environ. Res. Lett. 3, 024001 (6pp), 2008.

Smart, D. F. and M.A. Shea. Geomagnetic Cutoff Rigidity Calculations at 50-Year intervals between 1600 and 2000, Proc. 28th ICRC, Tsukuba, Japan, 4201-4204, 2003.

Smart, D.F. and M.A. Shea. Fifty years of progress in geomagnetic cutoff rigidity determinations. Advances in Space Research, 44, 1107-1123, 2009.

Smith, D. M., L. I. Lopez, R. P. Lin, and C. P. Barrington-Leigh. Terrestrial gamma-ray flashes observed up to 20 MeV, Science, 307 (5712), 1085–1088, doi:10.1126/science.1107466, 2005.

Song, P., Howard J. Singer and George L. Siscoe, editors. Space Weather. AGU Monograph Series, 125, pp. 440, Washington, DC, 2001.

Spurný, F., and Ts. Dachev. Intense Solar Flare Measurements, April 15, 2001, Radiat. Prot. Dosim., 95, p. 273-275, 2001.

Spurný, F., Kudela, K., Dachev, T. Airplane radiation dose decrease during a strong Forbush decrease. Space Weather 2, S05001, 2004.

Stoker, P. The IGY and beyond: A brief history of ground-based cosmic-ray detectors. Advances in Space Research, Volume 44, Issue 10, 1081-1095, 2009.

Storini, M. Galactic cosmic-ray modulation and solar-terrestrial relationships. Nuovo Cimento C, Serie 1, vol. 13 C, 103-124,1990.

Storini, M., G.A. Bazilevskaya, E.O. Flückiger, M.B. Krainev, V.S. Makhmutov and A.I. Sladkova, The Gnevyshev gap: A review for space weather, Adv. Space Res., 31, 4, 895-900, 2003.

Stozhkov, Y.I., N. S. Svirzhevsky, G. A. Bazilevskaya, M. B. Krainev, A. K. Svirzhevskaya, V. S. Makhmutov, V. I. Logachev, and E. V. Vashenyuk. Cosmic rays in the stratosphere in 2008–2010. Astrophys. Space Sci. Trans., 7, 379-382, 2011.

Strauss, R.D., M.S. Potgieter and S.E.S. Ferreira. Modeling Ground and Space Based Cosmic Ray Observations, Advances in Space Research, accepted manuscript, online, doi:10.1016/j.asr.2011.10.006, 2011.

Svensmark, H. and E. Friis-Christensen, Variation of cosmic ray flux and global cloud coverage-a missing link in solar-climate relationships, J. Atmos. Sol. Terr. Phys., 59, 1225–1232, 1997.

Thompson, J.L. Solar Diurnal Variation of Cosmic-Ray Intensity as a Function of Latitude. Phys. Rev., 54, 2, 93-96, 1938.

Tyasto, M.I., O. A. Danilova and V. E. Sdobnov. Variations in the Geomagnetic Cutoff Rigidity of CR in the Period of Magnetospheric Disturbances of May 2005: Their Correlation with Interplanetary Parameters, Bull. Russian Acad. Sci., ser. Phys., 75, 6, 808-811, 2011.

Usoskin, I.G., K. Alanko-Huotari, G.A. Kovaltsov and K. Mursula. Heliospheric modulation of cosmic rays: Monthly reconstruction for 1951-2004, JGR, 110, A12, A12108, 2005.

Usoskin, I.G., K. Mursula, S.K. Solanki, M. Schüssler, M. and G.A. Kovaltsov. A physical reconstruction of cosmic ray intensity since 1610. JGR, 107, A11, 1374, 2002.

Vainio, R., L. Desorgher, D. Heynderickx, M. Storini, E. O. Flückiger, R.B. Horne, G.A. Kovaltsov, K. Kudela, M. Laurenza, S. McKenna-Lawlor, H. Rothkaehl and I. Usoskin. Dynamics of the Earth's Particle Radiation Environment, Space Science Rev., 147, no. 3-4, 187-231. 2009.

Valdés-Galicia, J. F., et al., The Cosmic-Ray 1.68-Year Variation: a Clue to Understand the Nature of the Solar Cycle? Sol. Phys., 67, 409 – 417, 1996.

Valdés-Galicia, J.F. Energetic particle transport coefficients in the heliosphere, Space Science Reviews, 62, no. 1-2, 67-93, 1993.

Valdés-Galicia, J.F., Low energy galactic cosmic rays in the heliosphere, Adv. Space Res., 35, 755-767, 2005.

Van Allen, James A., Editor. Cosmic Rays, the Sun and Geomagnetism: The works of Scott E. Forbush, AGU, pp. 471, 1993.

Vashenyuk, E.V., Yu. V. Balabin, and B. B. Gvozdevsky. Features of relativistic solar proton spectra derived from ground level enhancement events (GLE) modeling, Astrophys. Space Sci. Trans., 7, 459-463, 2011.

Venkatesan, D. and Badruddin. Cosmic ray intensity variations in the 3-dimensional heliosphere, Space Sci. Rev., 52, 121-194, 1990.

Vilmer, N. A. L. MacKinnon, and G. J. Hurford. Properties of Energetic Ions in the Solar Atmosphere from γ-Ray and Neutron Observations, Space Sci. Rev., 159:167–224, 2011.

Vivek Gupta and Badruddin, Solar magnetic cycle dependence in corotating modulation of galactic cosmic rays, Astrophys Space Sci., 321: 185–195, 2009.

Wang, F. and Agrawal, V.D. Soft Error Considerations for Computer Web Servers, Proc. 42nd Southeastern Symposium on System Theory (SSST), 2010.

Webber, J.W. and J.A. Lockwood. Characteristics of the 22-year modulation of cosmic rays as seen by neutron monitors. JGR, 93, 8735-8740, 1988.

Zank, G.P. W.H. Matthaeus, J.W. Bieber and H. Moraal, The radial and latitudinal dependence of the cosmic ray diffusion tensor in the heliosphere, J. Geophys. Res., 103, A2, 2085-2097, 1998.

Zarrouk, N. and R. Bennaceur, Extrapolating cosmic ray variations and impacts on life: Morlet wavelet analysis, Acta Astronautica, 65, 1-2, 262-272, 2009.

Part 2

The Interaction of the
Solar Wind with the Magnetosphere

3

Impact of Solar Wind on the Earth Magnetosphere: Recent Progress in the Modeling of Ring Current and Radiation Belts

Natalia Buzulukova[1], Mei-Ching Fok[2] and Alex Glocer[2]
[1]*NASA Goddard Space Flight Center/CRESST/University of Maryland College Park*
[2]*NASA Goddard Space Flight Center*
USA

1. Introduction

When solar wind interacts with the Earth's magnetosphere it causes disturbances in the near-Earth plasma environment. Large disturbances result in geomagnetic storms, and affect not only Earth magnetosphere but also space-borne and ground-based technological systems. The systematic studies of cause-effect relations between solar wind variations and resulting disturbances in near-Earth plasma environment as well as the construction of relevant numerical models are important subjects of *Space Weather*, which is currently a very active topic of research.

A most common reaction of the magnetosphere to prolonged solar wind disturbances are geomagnetic storms. Understanding geomagnetic storms and their solar wind drivers is one of the most important problems in geophysics and space weather. A geomagnetic storms is defined by a deviation of the H component of the Earth's magnetic field at low latitudes, i.e., the Dst index. It is believed that the main source of this deviation is the so called ring current which is comprised of plasma with energies 1–200 keV. Hence, to understand geomagnetic storms, we need to understand how the ring current is formed. Many previous studies have addressed the question of solar wind drivers for geomagnetic storms. Most of these were statistical in nature (Borovsky & Denton, 2006; Denton et al., 2006; Turner et al., 2009; Weigel, 2010; Yermolaev et al., 2010; Zhang, Richardson, Webb, Gopalswamy, Huttunen, Kasper, Nitta, Poomvises, Thompson, Wu, Yashiro & Zhukov, 2007). They focused upon the effectiveness of different types of solar wind structures in producing geomagnetic disturbances (e.g., a combinations of global activity indices and/or various coupling functions) and different types of storms.

In addition to statistical analysis, a great deal of effort has been put into the development of ring current models. Typically a kinetic approach is used in which modeling ring current plasma is described with a number of species. Each population has its own energy and drift velocity which is a sum of the gradient/curvature drift and E×B drift. Plasma in this representation is generally anisotropic and is not in a thermodynamic equilibrium. The corresponding transport equations are usually written in terms of bounce-averaged quantities. This approach is sometimes referred to as 'drift physics' approach and has proven to be very useful in describing the plasma population of inner magnetosphere with energies from tens of eV to hundreds keV (e.g., Wolf et al. (2007)). Ring current models can also incorporate losses due to loss cone, charge exchange, Coulomb interactions, and interactions between different species due to wave activity. (For a detailed description of processes which control ring current losses, see Ebihara & Ejiri (2003) and Khazanov (2011)).

The electromagnetic field in the simulation region can be decomposed into a sum of 'external' field and 'internal' fields. The 'external' field is created outside of the simulation region while the 'internal' field is created by the ring current itself. For the sake of simplicity, the internal contribution is sometimes omitted. There exist a number of different codes which solve this problem with different assumptions, equations, and numerical methods (e.g., Ebihara & Ejiri (2003)) and many of them are widely used by the geophysical community. The Rice Convection Model (RCM) calculates internal electric fields self-consistently with the total pressure distribution (Harel et al., 1981; Jaggi & Wolf, 1973; Toffoletto et al., 2003). An externally prescribed magnetic field is assumed in the RCM. Each species is described by one adiabatic invariant (energy invariant), and the particle distribution function is assumed to be isotropic in pitch-angle. Self-consistency of the electric field with the plasma distribution is maintained by feedback from the ionosphere via field-aligned currents.

The Fok Ring Current (FokRC) model (Fok & Moore, 1997; Fok et al., 1995) and the Ring Current-Atmosphere Interaction Model (RAM) (Jordanova et al., 1994; 1997; Liemohn et al., 1999) solves the bounce-averaged Boltzmann equation for a number of species with given 'external' electric and magnetic fields. Each species is described by two adiabatic invariants μ, K (FokRC model) or, equivalently, energy and equatorial pitch angle (RAM model). The anisotropic pitch angle dependence of distribution function is calculated in both models.

The Comprehensive Ring Current Model (CRCM) (Fok, Wolf, Spiro & Moore, 2001) and the Self-Consistent RAM model (Ridley & Liemohn, 2002) calculate the internal electric fields self-consistently. The equilibrium RCM, RCM-E [Lemon et al., 2004] is a modification of RCM model with internal ring current magnetic field model that is calculated from 3-D force equilibrium assumption and treated as a correction to an external empirical magnetic field model. A modification of the RAM model with an internal ring current magnetic field also uses a 3-D force equilibrium assumption to calculate corrections to an external empirical magnetic field model (Zaharia et al., 2006). The Enhanced CRCM model (ECRCM) (Ebihara et al., 2008) is a modification of CRCM model with an internal ring current magnetic field that is calculated from the Biot-Savart law and included as a correction to an external empirical magnetic field model.

In addition to models of ring current plasma, there has been an increased interest in recent years in models of Earth's radiation belts. The Earth's radiation belts consist of energetic electrons (~100 keV to several MeV) and ions (~100 keV to several hundred MeV) trapped in the magnetosphere roughly from $1.2 < L < 8$. The energetic electrons reside in 2 distinct regions: the inner belt and the outer belt, which are usually separated by the slot region ($1.8 < L < 3$) of depleted particle populations. Pitch-angle diffusion loss of electrons by inter-acting with whistler mode plasmaspheric hiss is believed to be the cause of the slot region (Albert, 1994; Lyons et al., 1972; Meredith et al., 2007). The inner belt is relatively stable while the outer belt is highly variable with geomagnetic activity.

High energy electrons of radiation belts constitute a potential danger for Earth-orbiting satellites. High fluxes of energetic electrons cause harmful charging effects for sensitive electronics (Baker, 2002). The ability to predict how the radiation belts will react to solar wind drivers, and to geomagnetic storms in particular, is extremely important both for space weather applications and for geophysics in general. Approximately half of all moderate and intense storms cause a net increase of radiation belt fluxes by a factor of 2 or more (Reeves et al., 2003). At the same time, a quarter of all storms result in a net decrease of the fluxes by more than a factor of 2 (Reeves et al., 2003). There is an interplay between the number of processes that define the response of radiation belts to solar wind driving: acceleration, transport (convective and diffusive) and losses (Elkington et al., 1999; Fälthammar, 1965; Friedel et al., 2002; Horne, 2007; Schulz & Lanzerotti, 1974; Shprits et al., 2008; Summers et al., 2007; Ukhorskiy & Sitnov, 2008).

The main challenge to the community is how to describe and predict the variability of the radiation belts. There are three main approaches: (1) statistical approach based on analysis of relations between radiation belt fluxes and solar wind parameters like velocity and density, e.g. (Lyatsky & Khazanov, 2008; Reeves et al., 2011) and references therein; (2) modeling of electron phase space density with Fokker-Plank type equation (e.g. (Albert & Young, 2005; Li et al., 2001; Shprits et al., 2009; Varotsou et al., 2005)); (3) modeling of electron phase space density with combined Fokker-Plank and drift advection equation (Bourdarie et al., 1997; Fok et al., 2008; Miyoshi et al., 2006; Zheng et al., 2003). A necessity including advection terms (for gradient B drift and ExB drift) is dictated by the extension of a model to wider energy and L-shell ranges. In this case, a radiation belt model is similar to a ring current model with additional wave-diffusion terms. As a result both of these populations of near-Earth plasma, ring current and radiation belts, may be described within the same numerical model. This approach enables study of the complex and uneven relations between ring current and radiation belts. For example, when the ring current changes the electric field in the inner magnetosphere it influences radiation belts in (at least) two different ways: First, the electric field is important for convective transport. Second, electric field shapes the plasmasphere, a reservoir of cold and dense plasma of ionospheric origin. The plasmasphere directly drives radiation belt dynamics by excitation of waves; wave-particle interaction are an important acceleration mechanism for radiation belt electrons. Hence, we should describe the radiation belt dynamics together with the ring current and plasmasphere to ensure we do not miss important physics.

In this review paper, we will describe recent progress in the modeling of the ring current/radiation belt plasma with the Comprehensive Ring Current Model (CRCM) (Fok, Wolf, Spiro & Moore, 2001) and Radiation Belt Environment model (RBE) (Fok et al., 2005; Fok, Moore & Spjeldvik, 2001; Fok et al., 2008). Both models are based on a solution of the bounce-averaged transport equation for evolution of phase space density. We will describe the current progress of coupling the CRCM and the RBE with global MHD model of the Earth magnetosphere, modeling of ring current-plasmasphere interactions with the CRCM, and recent extensions to the RBE. In the discussion, we will describe an efforts we are going to make toward a combined ring current-radiation belt-plasmasphere model and putting this model inside a global MHD. The overall objective of this research can be described as follows: we are going to create a fully coupled model of ring current, radiation belts, plasmasphere and MHD with a significant level of self-consistency between all elements. This will be an excellent tool to study complex and non-linear response of Earth's inner magnetosphere to solar wind drivers.

2. Description of the Comprehensive Ring Current Model (CRCM)

The CRCM solves the distributions of ring current ions and electric potential at the ionosphere in a self-consistent manner (Fok, Wolf, Spiro & Moore, 2001). The temporal variation of the phase space density of ring current species is calculated by solving the bounce-averaged transport equation (Fok & Moore, 1997; Fok et al., 1996):

$$\frac{\partial \bar{f}_s}{\partial t} + \langle \dot{\lambda} \rangle \frac{\partial \bar{f}_s}{\partial \lambda} + \langle \dot{\phi} \rangle \frac{\partial \bar{f}_s}{\partial \phi} = -v\sigma_s \langle n_H \rangle \bar{f}_s - \left(\frac{\bar{f}_s}{0.5\tau_b} \right)_{loss\,cone} \tag{1}$$

where $\bar{f}_s = \bar{f}_s(\lambda, \phi, M, K)$ is the average distribution function of species s on the field line between mirror points. λ and ϕ are the magnetic latitude and local time, respectively, at the ionospheric foot point of the geomagnetic field line. M is the relativistic magnetic moment and $K = J/\sqrt{8m_0 M}$, where J is the second adiabatic invariant. The motion of the particles is described by their drifts across field lines, which are labeled by their ionospheric foot points. The M range is chosen to cover the energy ranges from 1–200 keV. The K range is chosen tending to cover the loss cone so that particle precipitations can be calculated.

The left hand side of (1) represents the drifts of the particle population. The right hand side of (1) refers to losses. The calculation of the bounce-averaged drift velocities, $\langle \dot{\lambda} \rangle$ and $\langle \dot{\phi} \rangle$, are described in detail in Fok & Moore (1997). $\langle \dot{\lambda} \rangle$ represents the radial transport term and $\langle \dot{\phi} \rangle$ is the azimuthal drift velocity. These drifts include gradient-curvature drift and E × B drift from convection and corotation electric fields. The effects of inductive electric fields due to time-varying magnetic fields are also taken into account implicitly in the model. For this purpose, we assume that field lines are rooted at the ionosphere, so that the inductive electric field there is zero. However, the shapes of field lines at higher altitudes vary as a function of time according to the magnetic field model. If field lines are perfect conductors, the field line motion at high altitudes (e.g., at the equator) will generate an induction electric field with the form, $E_{ind} = -v_0 \times B_0$ where B_0 and v_0 are the field line velocity and magnetic field at the equator.

The right hand side of (1) includes two loss terms; one is for the loss-cone and the other for the charge-exchange. To solve Eq. (1), we specify the particle distribution on the nightside boundary, which is set at 8–10 R_E or at the last closed field line, which one comes first. We use the Tsyganenko & Mukai (2003) model for density/temperature at the polar boundary, or output from the BATSRUS MHD (two-way coupled with ring current model). To describe O^+ component, we use the standard relation of Young et al. (1982).

The ionospheric electric field is calculated self-consistently with the ring current distribution. Field-aligned currents are calculated from a current continuity equation between magnetosphere and ionosphere (Fok, Wolf, Spiro & Moore, 2001):

$$J_{\|i} = \frac{1}{r_i \cos^2 \lambda} \sum_j \left(\frac{\partial \eta_j}{\partial \lambda} \frac{\partial W_j}{\partial \phi} - \frac{\partial \eta_j}{\partial \phi} \frac{\partial W_j}{\partial \lambda} \right) \tag{2}$$

where summation is at a fixed (λ, ϕ) point and over all (M, K) points; $J_{\|i}$ is a sum of ionospheric field-aligned current density for two hemispheres; W_j is the kinetic energy of a particle with given (λ, ϕ, M, K); and, η_j is the number of particles per unit magnetic flux associated with ($\Delta M, \Delta K$): $\eta_j = 4\sqrt{2}\pi m_0^{3/2} \bar{f}_s(\lambda, \phi, M, K)M^{1/2}\Delta M\Delta K$. Using the distribution of field-aligned currents, the ionospheric potential is obtained from $\nabla \cdot (-\Sigma \cdot \nabla\Phi) = J_{\|i} \sin I$, where Φ is an electric field potential (the same for both hemispheres by definition); Σ is a tensor of ionospheric conductivities (field-line integrated, a sum for two hemispheres); and, I is a magnetic field inclination angle. The CRCM grid is assumed to be the same in both hemispheres. This is usually true with the exception of points near polar CRCM boundary (e.g. Buzulukova et al. (2010)). Assuming the source of field-aligned current is located well inside the polar boundary, we may expect that errors in the solution for the potential are not too large. The detailed analysis of the asymmetry problem, however, is beyond the scope of this work and needs to be addressed separately.

3. Coupling of the CRCM with BATSRUS MHD

A reason for coupling of a ring current model with a global MHD model is twofold. First, MHD models apparently fail to adequately describe the inner magnetosphere. It follows from MHD assumption that the E×B drift velocity is dominant. This assumption is not valid in the inner magnetosphere where magnetic fields and their gradients are strong. Under such conditions, plasma cannot be treated as single fluid with some flow velocity and temperature. Hence, a coupling between a ring current model and MHD model provides MHD the missing 'drift physics' for better representation of the inner magnetosphere. Second, ring current models need a self-consistent input. Usually, ring current models use several empirical models as input. There is no guarantee these empirical models will be self-consistent with each other. MHD model can provide a ring current model with fully self-consistent electromagnetic fields and plasma parameters.

Here we present the results of so-called 'one-way coupling' between the CRCM and the Block-Adaptive-Tree Solar-Wind Roe-Type Upwind Scheme (BATSRUS) MHD model (Powell et al., 1999). An approach we follow is similar to one described in (De Zeeuw et al., 2004; Toffoletto et al., 2004). A computational domain is divided into two regions: 'inner' magnetosphere and 'outer' magnetosphere. The boundary between these two domains is located near the open/closed magnetic field lines boundary. In the outer magnetosphere, only the MHD model is calculated. In the inner magnetosphere, the MHD model runs together with RC model. In 'two-way' coupling mode the pressure from ring current model is mapped back to the MHD grid providing a feedback. This approach was called two-way coupling to emphasize that ring current and MHD models exchange information between each other. A first step is to couple the CRCM and BATSRUS in 'one-way' mode without pressure feedback from ring current to MHD. It means BATSRUS model provides the CRCM with an 'external' B field, electric field potential near the polar cap, and plasma pressure / temperature at the boundary between the two domains (open/closed boundary). The CRCM calculates the particle distribution function and 'internal' subauroral electric field in the inner magnetosphere. Toffoletto et al. (2004) considered 'one-way' coupling of the RCM with the Lyon-Fedder-Mobarry (LFM) MHD code. De Zeeuw et al. (2004) considered 'two-way' coupling of the RCM with BATSRUS MHD code. Only quasistationary states were considered and a substorm due to northward-southward Bz turning was not included.

For the coupled CRCM-BATSRUS, we show here the results for the idealized case of southward-northward-southward Bz turning and substorm development. The details of coupling methodology and run setup can be found in Buzulukova et al. (2010). Fig.1 from Buzulukova et al. (2010) shows the Cross Polar Cap Potential (CPCP) and IMF Bz for the MHD BATSRUS run.

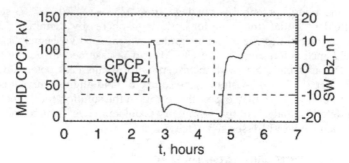

Fig. 1. MHD BATSRUS cross polar cap potential (CPCP) and solar wind Bz for an idealized case. Adopted from Buzulukova et al. (2010).

There is a characteristic structure of CPCP in response to the variation of Bz. A quasistationary CPCP ~110 kV at southward Bz corresponds to a strong convection. After the CPCP reaches the minimum at t = 3 h, northward IMF-associated (NBZ) current system (Iijima & Shibaji, 1987; Rasmussen & Schunk, 1987) begins to develop. At t~4.6 h, a southward Bz arrives at the magnetopause and cancels the NBZ current system. A typical two-cell convection system is restored. The substorm growth phase begins at ~5 h followed by substorm onset a t~5.3 h. At the end of the simulation, the profile reaches a plateau of 110 kV, the same as at the beginning of the run.

Fig.2 shows the dynamics of the inner magnetosphere for several snapshots during the CRCM-BATSRUS run. For each selected time, a plot of total RC pressure and a plot of field-aligned current are shown. All quantities are mapped to the equatorial plane. The plot of field-aligned currents is overlayed with equipotentials for the convection electric field without corotation. At the end of the Bz southward interval (t = 2.57 h), the RC is strongly asymmetric. There is a band of enhanced electric field in the dusk sector (MLT ~20) which can be associated with a Subauroral Ion Drift/Polarization Jet signature (SAID/PJ) (Galperin et al., 1974; Spiro et al., 1978). This feature is formed after ~1 h of simulations and can be interpreted as a result of shielding from enhanced electric field (Southwood & Wolf, 1978). Shielding is also manifested by the skewing of the equipotential lines in the morning sector (Brandt et al., 2002; Wolf et al., 2007). At t = 2.93 h, the polar cap potential drops substantially, and the RC-imposed electric field becomes apparent. This is an overshielding effect. At t = 4.64 (the end of the northward Bz interval) the RC becomes symmetric, as expected, and the RC-imposed electric field is almost zero.

At t = 4.7 h, the potential drop begins to increase; the RC becomes asymmetric again. As a result, a strong longitudinal pressure gradient and field-aligned current arises at t = 5.1 h. In the dusk sector (MLT ~20), field-aligned currents trigger the appearance of strong electric fields that resemble a SAID/PJ signature. This structure intensifies in the end of the growth phase. This is an effect of RC reconfiguration. The pressure gradients are formed by the preexisting population, so the time of formation is short, about ~15 min after the potential drop starts to increase. After the dipolarization starts at t=5.3 h, a strong convection and an induction electric field push particles into the inner magnetosphere, creating an injection. The injection is concentrated on the nightside and makes a quadrupole structure of field-aligned currents (t = 5.4, 5.53 h).

At t = 5.53 h, two populations persist: a preexisting one, and a newly injected population. Each population is spatially confined, and carries two layers of field-aligned currents. This layered structure is clearly seen in the dusk sector and interlaced in the dawn sector where the two populations coexist. There are two separated signatures of SAID/PJ in the dusk sector: one of them corresponds to the old population and the second corresponds to the newly injected population. Gradually the new population merges with the old one. At t = 6.8 h, a new quasi-stationary picture is established, similar to that at t = 2.5 h before the southward-northward Bz turning, with quasistationary SAID/PJ signature, shielding, and Region II currents.

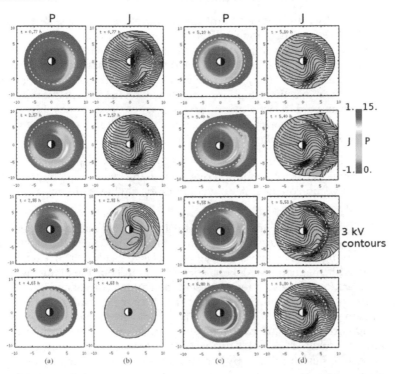

Fig. 2. Ring current dynamics for the CRCM-BATSRUS run. (a and c) Ring current (RC) pressure for 1–180 H+ (in color, logarithmic scale) at different times, t. (b and d) Snapshots of ionospheric Birkeland currents (BC) for the same times (in color) overplotted with convection equipotentials; corotation is excluded. All quantities are mapped onto the equatorial plane. BC and RC pressure are in mA/m² (from -1 to +1) and nPa (from 0 to 15), respectively. Equipotentials are 3 kV spacing. BC are positive down into the ionosphere. Sun is to the left. Dashed circles denote geostationary orbit. Adopted from Buzulukova et al. (2010).

For this idealized event the CRCM-BATSRUS coupled model successfully reproduces well known features of inner magnetosphere electrodynamics: strong convection under the southward Bz; the development of an electric field shielding and overshielding; a weak convection under prolonged interval of northward Bz; an induction electric field during substorm development; SAID/PJ signatures in the dusk sector during strong convection. Moreover, the stable structure of Region II field-aligned currents is formed only during prolonged intervals of strong quasi-stationary convection. During transient states when Bz changes polarity and/or substorm occurs, multilayered structures of field-aligned currents are formed. We interpret these as complex structures of partial ring current (including injections), created by spatial and temporal variations of electric fields (both convection and induction), and plasma sheet source at polar boundary of ring current model. In the next section we consider another case which demonstrate a complex structure of partial ring current and field-aligned currents during transient states, and associated plasmapause undulations.

4. CRCM with plasmasphere model and plasmapause undulations

On 17 April 2002, IMAGE EUV imager captured clear images of plasmaspheric undulations (Goldstein et al., 2005). They found shortly after a substorm onset at 1900 UT, the plasmasphere ripple was propagating westward along the plasmapause. Fok et al. (2005) used the combined models of the CRCM and the dynamic plasmasphere model of Ober et al. (1997) to simulate this event. They were able to reproduce the observed undulation and suggested that substorm injection, shielding and convection fields all play a role. We revisited this problem and examined in detail the effects of magnetosphere-ionosphere coupling and inductive electric field on the formation of plasmaspheric undulation with the CRCM model and Ober et al. (1997) model of plasmasphere (Buzulukova et al., 2008).

The dynamic plasmasphere model of Ober et al. (1997) calculates the plasma flux tube contents and equatorial plasma density distribution over time in the inner magnetosphere. It includes the influences of convection on the flux tube volumes; the daytime refilling; and, the nighttime draining of plasma.

Columns 1–3 in Fig. 3 show the simulated potential pattern, field aligned current mapped to the equator, plasmasphere density and ring current pressure at 3 times during the undulation event in a time-varying magnetic field of Tsyganenko (Tsyganenko, 1995). Column 4 displays results at t = t0 with magnetic field kept constant throughout the simulation. In Fig. 3, field-aligned current (top panels) in red represents downward current and blue is for upward current. The arrows in the plasmasphere plots (middle panels) indicate the locations of the ripples. As shown in Fig. 3 at t = t0 (column 1), particle injection on the nightside forms a bulge in the ring current pressure near the geostationary orbit in the dusk-midnight sector. Thirty minutes later when cross polar cap potential decreases due to northward turning of IMF Bz, the shielding field produced by the strong field aligned current dominates over the convection electric field (overshielding). A reverse flow region is formed near dusk local time outside the plasmapause and a ripple at the plasmapause is developed westward of this special flow pattern (column 2, middle panel). The ripple then propagates westward as the ring current-field aligned current move westward (column 3). The westward edge of the ring current pressure bulge in column 1 is potentially interchange unstable following the criterion of Xing & Wolf (2007). It is possible that interchange instability enhances the development of undulations for this particular event. Column 4 shows the configuration at t = t0 if magnetic field is kept constant. It can be seen that the particle injection is weaker without the inductive electric field associated with magnetic field fluctuations. As a result, no plasmaspheric undulation is developed at later times (results not shown). This example demonstrates the complexity and nonlinearity of ring current-ionosphere-plasmasphere. It also shows the ring current's relations to the plasmasphere undulation, overshielding, induction electric field and interchange instability.

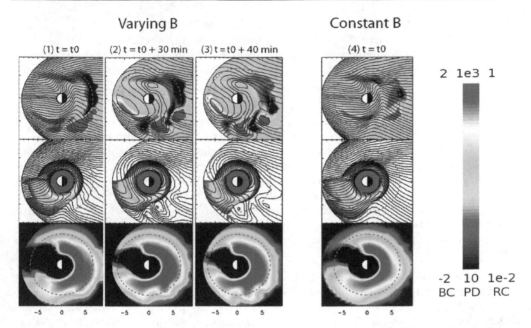

Fig. 3. Ring current-plasmasphere interaction. Columns 1-3 are simulated Region 2 Birkleland current (BC) mapped to the equator (top panels), plasmasphere density (PD) (middle panels) and ring current (RC) pressure for 16–27 keV H+ (bottom panels) at 3 times during the substorms on 17 April 2002. Column 4 is simulation result with constant magnetic field configuration. The dashed cycles are geosynchronous orbits. Potential contours without and with corotation are overlaid in top and middle panels, respectively. Adapted from Buzulukova et al. (2008).

5. The Radiation Belt Environment model (RBE)

The Radiation Belt Environment (RBE) model has been rigorously developed over past 10 years (Fok et al., 2005; Fok et al., 2011; Fok, Moore & Spjeldvik, 2001; Fok et al., 2008; Glocer et al., 2011; 2009; Zheng et al., 2011; 2003). It models both the radiation belts and ring current electrons. The RBE model is similar to the CRCM and is based on the solution of the bounce-averaged advection equation. Additional physics in the model includes the description of processes that do not conserve the first and the second adiabatic invariants via inclusion of diffusion for distribution functions. A radial transport is included implicitly by time-varying magnetic and electric fields, in the same manner as in the CRCM. Fok et al. (2008) consider only pitch-angle diffusion and energy diffusion. Recently, a mixed diffusion term has been added to the RBE model using the Alternating Direction Implicit (ADI) method (Fok et al., 2011; Zheng et al., 2011). The resulting equation is:

$$\frac{\partial \bar{f}_s}{\partial t} + \langle \dot{\lambda} \rangle \frac{\partial \bar{f}_s}{\partial \lambda} + \langle \dot{\phi} \rangle \frac{\partial \bar{f}_s}{\partial \phi} = \frac{1}{G} \frac{\partial}{\partial \alpha_0} \left[G \left(D_{\alpha_0 \alpha_0} \frac{\partial \bar{f}_s}{\partial \alpha_0} + D_{\alpha_0 E} \frac{\partial \bar{f}_s}{\partial E} \right) \right]$$

$$+ \frac{1}{G} \frac{\partial}{\partial E} \left[G \left(D_{EE} \frac{\partial \bar{f}_s}{\partial E} + D_{E\alpha_0} \frac{\partial \bar{f}_s}{\partial \alpha_0} \right) \right] - \left(\frac{\bar{f}_s}{0.5 \tau_b} \right)_{loss\,cone}$$

where $G = T(\alpha_0)\sin(2\alpha_0)(E + E_0)\sqrt{E(E + 2E_0)}$; E_0 is the rest mass energy; $T(\alpha_0) = 1/2R_0 \int_{s_m}^{s_m^*} ds / \cos\alpha$. To perform the diffusion in the (M, K) coordinates, we first map the particle phase space density from the (M, K) to (E, α_0) coordinates, perform diffusion in E, α_0, and then map the updated distribution back to the (M, K) coordinates (Fok et al., 1996).

The bounce-averaged diffusion coefficients are calculated with the Pitch-Angle and Energy Diffusion of Ions and Electrons (PADIE) code (Glauert & Horne, 2005). Corresponding normalized diffusion coefficients D_n is obtained that depend on $f = \omega_{pe}/\Omega_e$, the ratio of plasma frequency to the cyclotron frequency. To obtain cold plasma distribution and f we run Ober et al. (1997) model of plasmasphere. The corresponding diffusion coefficient is calculated by $D = D_n \cdot B_{wave}^2$ where B_{wave} is the wave intensity. Only resonance with lower-band whistler mode chorus is considered. The presence of chorus waves is confined between -15° and 15° magnetic latitude. The intensity of chorus waves is calculated from CRRESS plasma wave data for chorus presented by Meredith et al. (2001; 2003; 2009). The chorus wave amplitudes are based on an averaged study of chorus observations. However, the model has a radial dependence, an MLT dependence, and from a dependence on the level of geomagnetic activity.

A simplified version of the RBE model is currently running in real-time to provide radiation belt now-casting updated every 15min. The geosynchronous fluxes at longitudes of GOES-11 and 13 are extracted from the RBE real-time run and are plotted together with real-time GOES electron (>0.6 MeV) data. The model-data comparison is continually posted at http://mcf.gsfc.nasa.gov/RB_nowcast/.

Fok et al. (2011) study the effect of wave-particle interaction on radiation belt dynamics, including the effect of cross diffusion terms. Fig.4 shows simulated radiation belt fluxes on 23–27 October 2002 (adopted from (Fok et al., 2011, Fig.2)) as a function from L and t. Two energy bins are considered: low energy part of radiation belts (may be considered also as ring current electrons) 20–70 keV and high energy radiation belt electrons 0.6–1.8 MeV. Panels (a) and (b) shows electron fluxes without wave-particle interactions; panels (c) and (d) shows fluxes with pure energy/pitch angle diffusion; panels (e) and (f) shows fluxes with cross diffusion included. The Dst index (black curve) is overlaid on the top panels.

Radiation belt fluxes for different energies respond differently to storm conditions. For lower energies, ExB drift is important. During main phase of the storm, increased convection injects particles from higher L shells to lower L shells at the nightside part of the inner magnetosphere. Without wave-induced losses, freshly injected particles fill the entire outer belt (panel a). For MeV electrons, magnetic drifts dominate over convection. During main phase of strong storm, magnetic field in the inner magnetosphere becomes inflated because of ring current plasma. Inflation of magnetic field causes outward motion, de-acceleration and flux dropout of energetic particles. This is the well known Dst effect (Dessler & Karplus, 1961; Kim & Chan, 1997).

The inclusion of wave effects changes the results dramatically. Panels c and d show the results with pure energy and pitch-angle diffusion from interacting with whistler mode chorus for two energy bins. Panels e and f show the results with inclusion of cross diffusion coefficients.

For 20–70 keV, the inclusion of wave interactions significantly reduces radiation belt fluxes. It means pitch–angle diffusion is dominant. It scatters particles into loss cone causing precipitations. The difference in fluxes between panel (a) and panels (c) is the amount of

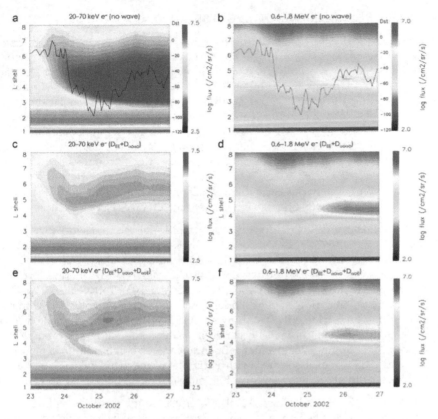

Fig. 4. Effect of wave-particle interaction on radiation belt dynamics. Simulated electron fluxes on 23–27 October 2002. Left panels: 20–70keV. Right panels: 0.6–18MeV. Top panels are fluxes without wave-particle interactions. Middle panels are fluxes with energy and pitch-angle diffusion. Bottom panels are fluxes with cross diffusion included. The black curves in the top panels are Dst. Adopted from Fok et al. (2011).

precipitated electron flux. For MeV electrons, the inclusion of chorus wave interactions causes gradual increase of fluxes during recovery phase of the storm. It means energy diffusion is dominant here.

Cross diffusion tends to moderate the impacts from pure pitch-angle and energy diffusion. For 20–70 keV, cross diffusion weakens pitch-angle diffusion and reduce precipitations. Overall flux calculated with cross diffusion is higher. For MeV electrons, cross-diffusion weakens energy diffusion and flux enhancement. Overall flux calculated with energy diffusion is lower than without it. The magnitude of cross diffusion effect depends from energy and L-shell. For MeV electrons at L-shell ~4, ignoring cross diffusion could cause overestimation of electron flux as much as a factor of 5. At times, cross diffusion may have a comparable effect to pure diffusions (Albert & Young, 2005). The cross-diffusion term can be especially important for ring current electrons with energies ~ 1 keV (Ni et al., 2011) therefore affecting electron precipitations.

6. Coupling the RBE model with MHD BATSRUS/SWMF

In most of previous RBE calculations, we used empirical models of magnetic and electric fields (Fok et al., 2005; Fok et al., 2008; Zheng et al., 2003). Variations in magnetic and electric fields are drivers for radial diffusion in RBE. Radial diffusion plays an important role in transport and energization of electrons in the outer belt. In this mechanism, electrons conserve the first and second adiabatic invariant, but the third invariant is regularly violated by temporal variations of the magnetic and induced electric field (Kellogg, 1959; Schulz & Eviatar, 1969). There is no explicit term for radial diffusion in RBE, rather this is handled implicitly through variations in the magnetic and electric fields. Therefore, a realistic description of these field variations is extremely important. This is the main motivation to couple the RBE model with global MHD model (Fok et al., 2011; Glocer et al., 2011; 2009). It is critical to use magnetic and electric fields from the two-way coupled MHD-ring current model for RBE, because ring current inflates inner magnetosphere magnetic field and changes electric field.

Currently, the RBE is placed inside the Space Weather Modeling Framework (SWMF) (Glocer et al., 2011; 2009). The SWMF consists of ~15 physics-based and several empirical models (modules) describing different processes in Sun-Earth coupled system (Tóth et al., 2005; Tóth et al., 2011). Inside the SWMF, the RBE is coupled with MHD BATSRUS, ring current model RCM and ionospheric solver RIM (for detail see Tóth et al. (2005); Tóth et al. (2011)). Coupling between MHD model and ring current model is done in two-way mode, i.e. with pressure feedback from ring current to MHD. It allows to include inflation of magnetic field during geomagnetic storms as well as realistic variation of magnetic field. For example, Huang, Spence, Hudson & Elkington (2010); Huang, Spence, Singer & Hughes (2010) demonstrate that MHD generates ULF waves associated with variations of magnetic field. These ULF waves are responsible for radial diffusion and result in diffusion rates similar to other estimations.

Recently Fok et al. (2011) and Glocer et al. (2011) use coupled RBE-SWMF to examine 2 events with rapid radiation belt enhancements. Based on careful data analysis from Akebono, TWINS and NOAA satellites, Glocer et al. (2011) conclude that the time scale of these enhancements is about 2 hours and thus it is too short for wave associated energization. Additional analysis of AL index and magnetic field measurements from GOES 11,12 satellites suggests a substorm may play a role in this quick increase of outer belt electrons (Fok et al., 2011; Glocer et al., 2011). The example of flux enhancement is shown on the Fig. 5 (adapted from Fok et al. (2011)). Panel (a) shows the Akebono electron flux of energy >2.5 MeV from September 3, 2008, 00 UT to September 4, 12 UT. The Dst index during this time is overlaid in the plot. The Akebono data are averaged over 3 orbit periods ($\sim$ 7.5 h).

Fok et al. (2011) and Glocer et al. (2011) perform RBE–SWMF simulations to understand the cause of the flux enhancement seen in the Akebono electron data. Additionally, the RBE is run in stand-alone mode with empirical models as inputs. Fig. 5(b) shows the RBE electron fluxes calculated with empirical T04 magnetic field model and empirical Weimer electric field model (Tsyganenko et al., 2003; Weimer, 2001). Fig. 5(c) is the RBE flux calculated in the MHD fields simulated from the RBE–SWMF model (De Zeeuw et al., 2004). When comparing the RBE fluxes with Akebono one should remember RBE fluxes are taken from equatorial plane with temporal resolution of 1 h. In contrast, the Akebono measurements are taken along high inclination orbits. Therefore the magnitudes may differ substantially as we are not sampling the same portion of the pitch-angle distribution. However, the temporal variability of high

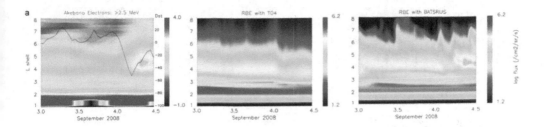

Fig. 5. Radiation belt flux enhancement. Left panel: L–Time plot of Akebono electron flux
(>2.5 MeV) on 3–4 September 2008. The black curve is Dst. Middle panel: corresponding
RBE simulated flux with T04 magnetic field. Right panel: simulated flux with BATSRUS
fields. Adopted from Fok et al. (2011).

latitude fluxes was found to be nearly identical with the equatorial fluxes (Kanekal et al.,
2001; 2005). Wave-particle interactions are not included in these RBE calculations. In the quiet
period on September 3, the two RBE simulations give similar flux intensity. During the main
phase of the storm, both RBE runs produce flux dropout in the outer belt. At 05–06 UT on
September 4, RBE-SWMF results show a sudden increase in electron flux consistent with the
Akebono data. There is no significant enhancement in the RBE–T04 run during the recovery
phase of the storm.

Since wave–particle interactions are not considered in these particular RBE calculations, the
enhancement seen in Fig. 5(c) has to be a result of particle transport. Glocer et al. (2011)
examine the magnetic configuration during the enhancement. It is found that the MHD model
predicts a substorm dipolarization at ~05 UT on September 4.

Fig. 6 shows the BATSRUS field lines in white and pressure in color on the X–Z plane in a
small time window surrounding the flux enhancement. Before the enhancement, the field in
the tail gets progressively more stretched. At 05:00 UT the stretching reaches a maximum
which coincides with the flux dropout. At 05:15 UT a reconnection site forms close to the
Earth and a plasmoid is ejected tailward. Magnetic reconnection which causes a substorm
development in this case is controlled by numerical dissipation, but for study of radiation belt
response to substorm, it suffices. As the dipolarization proceeds in the magnetic field, field
lines convect earthward. Electrons which are gyrating along these collapsing field lines can be
accelerated significantly on a time scale of minutes (Fok, Moore & Spjeldvik, 2001; Glocer et al.,
2009), much faster than the time scale for energization by whistler mode chorus waves, which
is typically of the order of 1-2 days (Summers & Ma, 2000). The T04 model, which is driven
by Dst and solar wind parameters, does not contain clear substorm signatures. Empirical
magnetic field models of this kind cannot directly simulate substorm reconfiguration unless
special tricks are applied (Delcourt et al., 1990; Fok, Moore & Spjeldvik, 2001; Pulkkinen et al.,
1991). For moderate storms such as this one on 3–5 September 2008, convection is weak and
the dominant energization and transport mechanism is sub-storm reconfiguration and the
resulting dipolarization electric field. The rapid enhancements of radiation belt fluxes during
modest storms cannot be explained without the consideration of substorm effects.

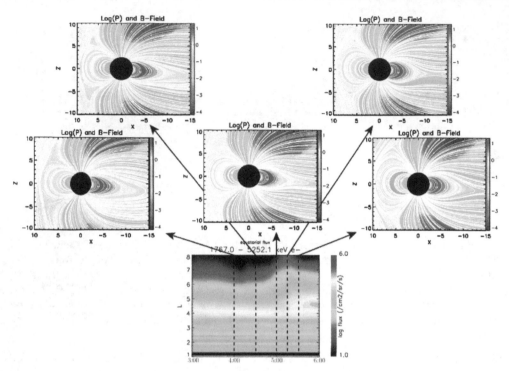

Fig. 6. MHD modeling of radiation belt enhancement and evolution of the magnetosphere. Top and middle panels: MHD plots of magnetic field lines (in white) and the pressure contours (color coded) in Y=0 plane. Bottom panel: The energetic electron flux from RBE from 03:00 to 06:00 UT on 4 September. Adopted from Glocer et al. (2011).

7. Discussion and directions for future work

The near-Earth space environment is described by multiscale physics that reflects a variety of processes and conditions occuring in magnetospheric plasma. Plasma densities vary from 10^6 cm^{-3} in the F layer of dayside ionosphere to less then 1 cm^{-3} in the tail. Ionospheric plasma is highly collisional, and the plasma above $\sim$500 km is essentially collisionless. It has been recently realized that for a successful description of such a plasma system, a complex solution is needed which allows different physics domains to be described using different physical models (De Zeeuw et al., 2004; Fok et al., 2006; Moore et al., 2008; Ridley et al., 2004; Tóth et al., 2005; Tóth et al., 2011; Zhang, Liemohn, De Zeeuw, Borovsky, Ridley, Toth, Sazykin, Thomsen, Kozyra, Gombosi & Wolf, 2007). This is the reason why coupled models are intensively developed and improved during recent years. Besides several cases considered here, the latest examples include: coupling between the RCM (ring current model) and Versatile Electron Radiation Belt (VERB) code (Subbotin et al., 2011); coupling between RAM-SCB code (ring current model) and BATSRUS/SWMF (MHD with the RCM) (Welling et al., 2011; Zaharia et al., 2010); coupling between two ring current models RAM and the RCM (Jordanova et al., 2010); one way coupling between MHD Open Geospace General Circulation Model (Open GGCM) – the RCM (ring current model) (Hu et al., 2010). None of these models however is 'ideal' (including the examples presented in detail earlier) in the sense that all of them have some shortcuts. For example, the RCM assumes an isotropic distribution function which is not consistent with a concept of diffusion in pitch-angle (RCM-VERB code); the RAM-RCM code does not contain a self-consistent magnetic field; the RAM-SCB-BATSRUS/SWMF as well as CRCM-BATSRUS and Open GGCM-RCM work only in one-way mode; coupled RBE-BATSRUS/SWMF code does not have wave interactions so far. To overcome at least some of these shortcuts, we are going to make the following improvements to CRCM/RBE/BATSRUS/SWMF model:

- Create two–way coupling between the CRCM and MHD BATSRUS;

- Combine the CRCM, the RBE and plasmasphere codes into one module of inner magnetosphere electrodynamics;

- Gradually include waves in addition to the chorus wave modes, such as plasmaspheric hiss and electron ion cyclotron waves into the RBE and the CRCM;

- Combine all modules (including two–way coupling between ring current and MHD) and incorporate them into the SWMF.

It should be noted that a coupling of different physical models is always nontrivial and some shortcuts are inevitable. However, each newly coupled model is an incremental improvement to the current state-of-the-art inner magnetosphere models. Each improvement enhances our overall understanding of complex and nonlinear relationships between different domains in the Earth's magnetosphere and the response of magnetosphere to solar wind drivers.

8. Acknowledgments

This research was supported by NASA Science Mission Directorate, Heliophysics Division, Living With a Star Targeted Research and Technology Program, under Work Breakdown Structures: 936723.02.01.06.78 and 936723.02.01.01.27, and Heliophysics Guest Investigators Program, under Work Breakdown Structure 955518. N. Buzulukova thanks E. Mitchell for proofreading of the manuscript.

9. References

Albert, J. M. (1994). Quasi-linear pitch angle diffusion coefficients: Retaining high harmonics, *J. Geophys. Res.* 992: 23741–+.

Albert, J. M. & Young, S. L. (2005). Multidimensional quasi-linear diffusion of radiation belt electrons, *Geophys. Res. Lett.* 321: L14110.

Baker, D. N. (2002). How to cope with space weather, *Science* 297(5586): 1486–1487.
    URL: *http://www.sciencemag.org/content/297/5586/1486.short*

Borovsky, J. E. & Denton, M. H. (2006). Differences between CME-driven storms and CIR-driven storms, *J. Geophys. Res.* 111: A07S08.

Bourdarie, S., Boscher, D., Beutier, T., Sauvaud, J.-A. & Blanc, M. (1997). Electron and proton radiation belt dynamic simulations during storm periods: A new asymmetric convection-diffusion model, *J. Geophys. Res.* 1021: 17541–17552.

Brandt, P. C., Ohtani, S., Mitchell, D. G., Fok, M.-C., Roelof, E. C. & Demajistre, R. (2002). Global ENA observations of the storm mainphase ring current: Implications for skewed electric fields in the inner magnetosphere, *Geophys. Res. Lett.* 29(20): 1359.

Buzulukova, N., Fok, M.-C., Moore, T. E. & Ober, D. M. (2008). Generation of plasmaspheric undulations, *Geophys. Res. Lett.* 351: L13105.

Buzulukova, N., Fok, M.-C., Pulkkinen, A., Kuznetsova, M., Moore, T. E., Glocer, A., Brandt, P. C., Tóth, G. & Rastätter, L. (2010). Dynamics of ring current and electric fields in the inner magnetosphere during disturbed periods: CRCM-BATS-R-US coupled model, *J. Geophys. Res.* 115: A05210.

De Zeeuw, D. L., Sazykin, S., Wolf, R. A., Gombosi, T. I., Ridley, A. J. & Tóth, G. (2004). Coupling of a global MHD code and an inner magnetospheric model: Initial results, *Journal of Geophysical Research (Space Physics)* 109: A12219.

Delcourt, D. C., Pedersen, A. & Sauvaud, J. A. (1990). Dynamics of single-particle orbits during substorm expansion phase, *J. Geophys. Res.* 952: 20853–20865.

Denton, M. H., Borovsky, J. E., Skoug, R. M., Thomsen, M. F., Lavraud, B., Henderson, M. G., McPherron, R. L., Zhang, J. C. & Liemohn, M. W. (2006). Geomagnetic storms driven by ICME- and CIR-dominated solar wind, *J. Geophys. Res.* 111: A07S07.

Dessler, A. J. & Karplus, R. (1961). Some Effects of Diamagnetic Ring Currents on Van Allen Radiation, *J. Geophys. Res.* 66: 2289–2295.

Ebihara, Y. & Ejiri, M. (2003). Numerical Simulation of the Ring Current: Review, *Space Sci. Rev* 105: 377–452.

Ebihara, Y., Fok, M.-C., Blake, J. B. & Fennell, J. F. (2008). Magnetic coupling of the ring current and the radiation belt, *J. Geophys. Res.* 113: 7221.

Elkington, S. R., Hudson, M. K. & Chan, A. A. (1999). Acceleration of relativistic electrons via drift-resonant interaction with toroidal-mode Pc-5 ULF oscillations, *Geophys. Res. Lett.* 26: 3273–3276.

Fälthammar, C.-G. (1965). Effects of Time-Dependent Electric Fields on Geomagnetically Trapped Radiation, *J. Geophys. Res.* 70: 2503–2516.

Fok, M. C., Ebihara, Y., Moore, T. E., Ober, D. M. & Keller, K. A. (2005). Geospace storm processes coupling the ring current, radiation belt and plasmasphere, *in* J. B. et al. (ed.), *Inner Magnetosphere Interactions: New Perspectives from Imaging*, Vol. 159 of *Geophys. Monogr. Ser.*, AGU, Washington, D.C., pp. 207–220.

Fok, M.-C., Glocer, A., Zheng, Q., Horne, R. B., Meredith, N. P., Albert, J. M. & Nagai, T. (2011). Recent developments in the radiation belt environment model, *J. Atmos. Sol. Terr. Phys.* 73: 1435–1443.

Fok, M.-C. & Moore, T. E. (1997). Ring current modeling in a realistic magnetic field configuration, *Geophys. Res. Lett.* 24: 1775–1778.

Fok, M.-C., Moore, T. E., Brandt, P. C., Delcourt, D. C., Slinker, S. P. & Fedder, J. A. (2006). Impulsive enhancements of oxygen ions during substorms, *J. Geophys. Res.* 111: 10222.

Fok, M. C., Moore, T. E., Kozyra, J. U., Ho, G. C. & Hamilton, D. C. (1995). Three-dimensional ring current decay model, *J. Geophys. Res.* 100: 9619–9632.

Fok, M.-C., Moore, T. E. & Spjeldvik, W. N. (2001). Rapid enhancement of radiation belt electron fluxes due to substorm dipolarization of the geomagnetic field, *J. Geophys. Res.* 106: 3873–3882.

Fok, M.-C., Wolf, R. A., Spiro, R. W. & Moore, T. E. (2001). Comprehensive computational model of Earth's ring current, *J. Geophys. Res.* 106: 8417–8424.

Fok, M., Horne, R. B., Meredith, N. P. & Glauert, S. A. (2008). Radiation Belt Environment model: Application to space weather nowcasting, *J. Geophys. Res.* 113: 3.

Fok, M., Moore, T. E. & Greenspan, M. E. (1996). Ring current development during storm main phase, *J. Geophys. Res.* 101: 15311–15322.

Friedel, R. H. W., Reeves, G. D. & Obara, T. (2002). Relativistic electron dynamics in the inner magnetosphere - a review, *Journal of Atmospheric and Solar-Terrestrial Physics* 64: 265–282.

Galperin, Y. I., Ponomarev, V. N. & Zosimova, A. G. (1974). Plasma convection in polar ionosphere, *Ann. Geophys.* 30(1): 1–7.

Glauert, S. A. & Horne, R. B. (2005). Calculation of pitch angle and energy diffusion coefficients with the PADIE code, *J. Geophys. Res.* 110: A04206.

Glocer, A., Fok, M.-C., Nagai, T., Tóth, G., Guild, T. & Blake, J. (2011). Rapid rebuilding of the outer radiation belt, *J. Geophys. Res.* 116: A09213.

Glocer, A., Toth, G., Fok, M., Gombosi, T. & Liemohn, M. (2009). Integration of the radiation belt environment model into the space weather modeling framework, *J. Atmos. Sol. Terr. Phys.* 71: 1653–1663.

Goldstein, J., Burch, J. L., Sandel, B. R., Mende, S. B., C:son Brandt, P. & Hairston, M. R. (2005). Coupled response of the inner magnetosphere and ionosphere on 17 April 2002, *Journal of Geophysical Research (Space Physics)* 110: A03205.

Harel, M., Wolf, R. A., Reiff, P. H., Spiro, R. W., Burke, W. J., Rich, F. J. & Smiddy, M. (1981). Quantitative simulation of a magnetospheric substorm. I - Model logic and overview, *J. Geophys. Res.* 86: 2217–2241.

Horne, R. B. (2007). Plasma astrophysics: Acceleration of killer electrons, *Nature Physics* 3: 590–591.

Hu, B., Toffoletto, F. R., Wolf, R. A., Sazykin, S., Raeder, J., Larson, D. & Vapirev, A. (2010). One-way coupled OpenGGCM/RCM simulation of the 23 March 2007 substorm event, *J. Geophys. Res.* 115: A12205.

Huang, C.-L., Spence, H. E., Hudson, M. K. & Elkington, S. R. (2010). Modeling radiation belt radial diffusion in ULF wave fields: 2. Estimating rates of radial diffusion using combined MHD and particle codes, *J. Geophys. Res.* 115: A06216.

Huang, C.-L., Spence, H. E., Singer, H. J. & Hughes, W. J. (2010). Modeling radiation belt radial diffusion in ULF wave fields: 1. Quantifying ULF wave power at geosynchronous orbit in observations and in global MHD model, *J. Geophys. Res.* 115: A06215.

Iijima, T. & Shibaji, T. (1987). Global characteristics of northward IMF-associated (NBZ) field-aligned currents, *J. Geophys. Res.* 92: 2408–2424.

Jaggi, R. K. & Wolf, R. A. (1973). Self-consistent calculation of the motion of a sheet of ions in the magnetosphere., *J. Geophys. Res.* 78: 2852–2866.

Jordanova, V. K., Kozyra, J. U., Khazanov, G. V., Nagy, A. F., Rasmussen, C. E. & Fok, M.-C. (1994). A bounce-averaged kinetic model of the ring current ion population, *Geophys. Res. Lett.* 21: 2785–2788.

Jordanova, V. K., Kozyra, J. U., Nagy, A. F. & Khazanov, G. V. (1997). Kinetic model of the ring current-atmosphere interactions, *J. Geophys. Res.* 102: 14279–14292.

Jordanova, V. K., Thorne, R. M., Li, W. & Miyoshi, Y. (2010). Excitation of whistler mode chorus from global ring current simulations, *J. Geophys. Res.* 115: A00F10.

Kanekal, S. G., Baker, D. N. & Blake, J. B. (2001). Multisatellite measurements of relativistic electrons: Global coherence, *J. Geophys. Res.* 106: 29721–29732.

Kanekal, S. G., Friedel, R. H. W., Reeves, G. D., Baker, D. N. & Blake, J. B. (2005). Relativistic electron events in 2002: Studies of pitch angle isotropization, *J. Geophys. Res.* 110: A12224.

Kellogg, P. J. (1959). Van Allen Radiation of Solar Origin, *Nature* 183: 1295–1297.

Khazanov, G. V. (2011). *Kinetic Theory of the Inner Magnetospheric Plasma*, Springer-Verlag.

Kim, H.-J. & Chan, A. A. (1997). Fully adiabatic changes in storm time relativistic electron fluxes, *J. Geophys. Res.* 1022: 22107–22116.

Li, X., Temerin, M., Baker, D. N., Reeves, G. D. & Larson, D. (2001). Quantitative prediction of radiation belt electrons at geostationary orbit based on solar wind measurements, *Geophys. Res. Lett.* 28: 1887–1890.

Liemohn, M. W., Kozyra, J. U., Jordanova, V. K., Khazanov, G. V., Thomsen, M. F. & Cayton, T. E. (1999). Analysis of early phase ring current recovery mechanisms during geomagnetic storms, *Geophys. Res. Lett* 26: 2845–2848.

Lyatsky, W. & Khazanov, G. V. (2008). Effect of geomagnetic disturbances and solar wind density on relativistic electrons at geostationary orbit, *J. Geophys. Res.* 113: A08224.

Lyons, L. R., Thorne, R. M. & Kennel, C. F. (1972). Pitch-angle diffusion of radiation belt electrons within the plasmasphere., *J. Geophys. Res.* 77: 3455–3474.

Meredith, N. P., Horne, R. B. & Anderson, R. R. (2001). Substorm dependence of chorus amplitudes: Implications for the acceleration of electrons to relativistic energies, *J. Geophys. Res.* 106: 13165–13178.

Meredith, N. P., Horne, R. B., Glauert, S. A. & Anderson, R. R. (2007). Slot region electron loss timescales due to plasmaspheric hiss and lightning-generated whistlers, *J. Geophys. Res.* 112: A08214.

Meredith, N. P., Horne, R. B., Thorne, R. M. & Anderson, R. R. (2003). Favored regions for chorus-driven electron acceleration to relativistic energies in the Earth's outer radiation belt, *Geophys. Res. Lett.* 30(16): 160000–1.

Meredith, N. P., Horne, R. B., Thorne, R. M. & Anderson, R. R. (2009). Survey of upper band chorus and ECH waves: Implications for the diffuse aurora, *J. Geophys. Res.* 114: A07218.

Miyoshi, Y. S., Jordanova, V. K., Morioka, A., Thomsen, M. F., Reeves, G. D., Evans, D. S. & Green, J. C. (2006). Observations and modeling of energetic electron dynamics during the October 2001 storm, *J. Geophys. Res.* 111: A11S02.

Moore, T. E., Fok, M.-C., Delcourt, D. C., Slinker, S. P. & Fedder, J. A. (2008). Plasma plume circulation and impact in an MHD substorm, *J. Geophys. Res.* 113: 6219.

Ni, B., Thorne, R. M., Meredith, N. P., Horne, R. B. & Shprits, Y. Y. (2011). Resonant scattering of plasma sheet electrons leading to diffuse auroral precipitation: 2. Evaluation for whistler mode chorus waves, *J. Geophys. Res.* 116: A04219.

Ober, D. M., Horwitz, J. L. & Gallagher, D. L. (1997). Formation of density troughs embedded in the outer plasmasphere by subauroral ion drift events, *J. Geophys. Res.* 102: 14595–14602.

Powell, K. G., Roe, P. L., Linde, T. J., Gombosi, T. I. & de Zeeuw, D. L. (1999). A Solution-Adaptive Upwind Scheme for Ideal Magnetohydrodynamics, *J. Comput. Phys.* 154: 284–309.

Pulkkinen, T. I., Baker, D. N., Fairfield, D. H., Pellinen, R. J., Murphree, J. S., Elphinstone, R. D., McPherron, R. L., Fennell, J. F., Lopez, R. E. & Nagai, T. (1991). Modeling the growth phase of a substorm using the Tsyganenko model and multi-spacecraft observations - CDAW-9, *Geophys. Res. Lett.* 18: 1963–1966.

Rasmussen, C. E. & Schunk, R. W. (1987). Ionospheric convection driven by NBZ currents, *J. Geophys. Res.* 92: 4491–4504.

Reeves, G. D., Henderson, M. C., Skoug, R. M., Thomsen, M. F., Borovsky, J. E., Funsten, H. O., C:Son Brandt, P., Mitchell, D. J., Jahn, J.-M., Pollock, C. J., McComas, D. J. & Mende, S. B. (2003). IMAGE, POLAR, and Geosynchronous Observations of Substorm and Ring Current Ion Injection, *in* A. S. Sharma, Y. Kamide, & G. S. Lakhina (ed.), *Disturbances in Geospace: The Storm-substorm Relationship*, Vol. 142 of *Washington DC American Geophysical Union Geophysical Monograph Series*, pp. 91–+.

Reeves, G. D., Morley, S. K., Friedel, R. H. W., Henderson, M. G., Cayton, T. E., Cunningham, G., Blake, J. B., Christensen, R. A. & Thomsen, D. (2011). On the relationship between relativistic electron flux and solar wind velocity: Paulikas and Blake revisited, *J. Geophys. Res.* 116: A02213.

Ridley, A., Gombosi, T. & Dezeeuw, D. (2004). Ionospheric control of the magnetosphere: conductance, *Ann. Geophys.* 22: 567–584.

Ridley, A. J. & Liemohn, M. W. (2002). A model-derived storm time asymmetric ring current driven electric field description, *J. Geophys. Res.* 107: 1151.

Schulz, M. & Eviatar, A. (1969). Diffusion of equatorial particles in the outer radiation zone., *J. Geophys. Res.* 74: 2182–2192.

Schulz, M. & Lanzerotti, L. J. (1974). *Particle diffusion in the radiation belts*, Springer-Verlag.

Shprits, Y. Y., Chen, L. & Thorne, R. M. (2009). Simulations of pitch angle scattering of relativistic electrons with MLT-dependent diffusion coefficients, *J. Geophys. Res.* 114: A03219.

Shprits, Y. Y., Elkington, S. R., Meredith, N. P. & Subbotin, D. A. (2008). Review of modeling of losses and sources of relativistic electrons in the outer radiation belt I: Radial transport, *Journal of Atmospheric and Solar-Terrestrial Physics* 70: 1679–1693.

Southwood, D. J. & Wolf, R. A. (1978). An assessment of the role of precipitation in magnetospheric convection, *J. Geophys. Res.* 83: 5227–5232.

Spiro, R. W., Heelis, R. A. & Hanson, W. B. (1978). Ion convection and the formation of the mid-latitude F region ionization trough, *J. Geophys. Res.* 83: 4255–4264.

Subbotin, D. A., Shprits, Y. Y., Gkioulidou, M., Lyons, L. R., Ni, B., Merkin, V. G., Toffoletto, F. R., Thorne, R. M., Horne, R. B. & Hudson, M. K. (2011). Simulation of the acceleration of relativistic electrons in the inner magnetosphere using RCM-VERB coupled codes, *J. Geophys. Res.* 116: A08211.

Summers, D. & Ma, C.-y. (2000). A model for generating relativistic electrons in the Earth's inner magnetosphere based on gyroresonant wave-particle interactions, *J. Geophys. Res.* 105: 2625–2640.

Summers, D., Ni, B. & Meredith, N. P. (2007). Timescales for radiation belt electron acceleration and loss due to resonant wave-particle interactions: 2. Evaluation for VLF chorus, ELF hiss, and electromagnetic ion cyclotron waves, *J. Geophys. Res.* 112: A04207.

Toffoletto, F. R., Sazykin, S., Spiro, R. W., Wolf, R. A. & Lyon, J. G. (2004). RCM meets LFM: initial results of one-way coupling, *J. Atmos. Sol. Terr. Phys.* 66: 1361–1370.

Toffoletto, F., Sazykin, S., Spiro, R. & Wolf, R. (2003). Inner magnetospheric modeling with the Rice Convection Model, *Space Sci. Rev.* 107: 175–196.

Tóth, G., Sokolov, I. V., Gombosi, T. I., Chesney, D. R., Clauer, C. R., De Zeeuw, D. L., Hansen, K. C., Kane, K. J., Manchester, W. B., Oehmke, R. C., Powell, K. G., Ridley, A. J., Roussev, I. I., Stout, Q. F., Volberg, O., Wolf, R. A., Sazykin, S., Chan, A., Yu, B. & Kóta, J. (2005). Space Weather Modeling Framework: A new tool for the space science community, *J. Geophys. Res.* 110: A12226.

Tóth, G., van der Holst, B., Sokolov, I. V., Zeeuw, D. L. D., Gombosi, T. I., Fang, F., Manchester, W. . B., Meng, X., Najib, D., Powell, K. G., Stout, Q. F., Glocer, A., Ma, Y.-J. & er, M. O. (2011). Adaptive numerical algorithms in space weather modeling, *J. Comput. Phys.,In Press, Corrected Proof* .

Tsyganenko, N. A. (1995). Modeling the Earth's magnetospheric magnetic field confined within a realistic magnetopause, *J. Geophys. Res.* 100: 5599–5612.

Tsyganenko, N. A. & Mukai, T. (2003). Tail plasma sheet models derived from Geotail particle data, *J. Geophys. Res.* 108: 1136.

Tsyganenko, N. A., Singer, H. J. & Kasper, J. C. (2003). Storm-time distortion of the inner magnetosphere: How severe can it get?, *J. Geophys. Res.* 108: 1209.

Turner, N. E., Cramer, W. D., Earles, S. K. & Emery, B. A. (2009). Geoefficiency and energy partitioning in CIR-driven and CME-driven storms, *J. Geophys. Res.* 71: 1023–1031.

Ukhorskiy, A. Y. & Sitnov, M. I. (2008). Radial transport in the outer radiation belt due to global magnetospheric compressions, *J. Atmos. Sol. Terr. Phys.* 70: 1714–1726.

Varotsou, A., Boscher, D., Bourdarie, S., Horne, R. B., Glauert, S. A. & Meredith, N. P. (2005). Simulation of the outer radiation belt electrons near geosynchronous orbit including both radial diffusion and resonant interaction with Whistler-mode chorus waves, *Geophys. Res. Lett.* 321: L19106.

Weigel, R. S. (2010). Solar wind density influence on geomagnetic storm intensity, *J. Geophys. Res.* 115: A09201.

Weimer, D. R. (2001). An improved model of ionospheric electric potentials including substorm perturbations and application to the Geospace Environment Modeling November 24, 1996, event, *J. Geophys. Res.* 106: 407–416.

Welling, D. T., Jordanova, V. K., Zaharia, S. G., Glocer, A. & Toth, G. (2011). The effects of dynamic ionospheric outflow on the ring current, *J. Geophys. Res.* 116: A00J19.

Wolf, R. A., Spiro, R. W., Sazykin, S. & Toffoletto, F. R. (2007). How the Earth's inner magnetosphere works: An evolving picture, *J. Atmos. Sol. Terr. Phys.* 69: 288–302.

Xing, X. & Wolf, R. A. (2007). Criterion for interchange instability in a plasma connected to a conducting ionosphere, *J. Geophys. Res.* 112: A12209.

Yermolaev, Y. I., Nikolaeva, N. S., Lodkina, I. G. & Yermolaev, M. Y. (2010). Specific interplanetary conditions for CIR-, Sheath-, and ICME-induced geomagnetic storms obtained by double superposed epoch analysis, *Annales Geophysicae* 28: 2177–2186.

Young, D. T., Balsiger, H. & Geiss, J. (1982). Correlations of magnetospheric ion composition with geomagnetic and solar activity, *J. Geophys. Res.* 87: 9077–9096.

Zaharia, S., Jordanova, V. K., Thomsen, M. F. & Reeves, G. D. (2006). Self-consistent modeling of magnetic fields and plasmas in the inner magnetosphere: Application to a geomagnetic storm, *J. Geophys. Res.* 111: 11.

Zaharia, S., Jordanova, V. K., Welling, D. & Tóth, G. (2010). Self-consistent inner magnetosphere simulation driven by a global MHD model, *J. Geophys. Res.* 115: A12228.

Zhang, J., Liemohn, M. W., De Zeeuw, D. L., Borovsky, J. E., Ridley, A. J., Toth, G., Sazykin, S., Thomsen, M. F., Kozyra, J. U., Gombosi, T. I. & Wolf, R. A. (2007). Understanding storm-time ring current development through data-model comparisons of a moderate storm, *J. Geophys. Res.* 112(A11): 4208.

Zhang, J., Richardson, I. G., Webb, D. F., Gopalswamy, N., Huttunen, E., Kasper, J. C., Nitta, N. V., Poomvises, W., Thompson, B. J., Wu, C.-C., Yashiro, S. & Zhukov, A. N. (2007). Solar and interplanetary sources of major geomagnetic storms (Dst < -100 nT) during 1996-2005, *J. Geophys. Res.* 112: A10102.

Zheng, Q., Fok, M.-C., Albert, J., Horne, R. B. & Meredith, N. P. (2011). Effects of energy and pitch angle mixed diffusion on radiation belt electrons, *J. Atmos. Sol. Terr. Phys.* 73: 785–795.

Zheng, Y., Fok, M.-C. & Khazanov, G. V. (2003). A radiation belt-ring current forecasting model, *Space Weather* 1: 1013.

4

The Polar Cap PC Indices: Relations to Solar Wind and Global Disturbances

Peter Stauning
Danish Meteorological Institute
Denmark

1. Introduction

The solar wind plasma flow impinging on the Earths magnetosphere causes a range of geophysical disturbances such as magnetic storms and substorms, energization of the plasma trapped in the Earth's near space, auroral activity, and Joule heating of the upper atmosphere. Auroral activity was characterized already in 1961 by the electrojet indices (AU, AL, AE and AO) based on magnetic observations at auroral latitudes (Davis & Sugiura, 1966). On the contrary, several attempts have been made in the past to scale high-latitude disturbances on basis of polar magnetic variations without reaching a final, agreed and IAGA-adopted Polar Cap index. Fairfield (1968) suggested that the maximum amplitude of the magnetic variations observed from a ring of polar cap observatories could be a better indicator of the overall high-latitude magnetic activity than the auroral electrojet indices. He also noted that the Polar Cap "magnetic activity magnitude" sometimes increased before changes in the AE index.

Kuznetsov & Troshichev (1977) defined a "PC_L" index based on the variability of high-latitude magnetic recordings (much like the K indices), and not equivalent to the present PC "level index". Later, a "PC(Bz)" index based on a composite of the variance and the level of polar magnetic activity was proposed by Troshichev et al. (1979), who used the Polar Cap magnetic activity as a signature of substorm development. The "MAGPC" index suggested by Troshichev & Andrezen (1985) was based on the magnitude (in nT) of 15 min samples of the magnetic variation in the direction of the 03:00-15:00 MLT meridian. The MAGPC index was introduced to provide a measure of the solar wind electric field to be derived from available ground-based magnetic observations in the central Polar Cap. A major problem for these initial "PC" indices was their dependence on the daily and seasonal changes in the ionospheric conductivity with the varying solar illumination. Such variations cause corresponding variations in the "sensitivity" of the disturbance indices in their response to varying solar wind conditions.

The present version of a Polar Cap (PC) index is based on the formulation by Troshichev et al. (1988). The new idea is the scaling on a statistical basis of the magnetic variations to the electric field in the solar wind in order to make the new PC index independent of the local ionospheric properties and their daily and seasonal variations. The PC index concept was further developed by Vennerstrøm (1991) and Vennerstrøm et al. (1991). The development of a PC index was recommended by IAGA in 1999 and the index was later adopted by IAGA as an international standard index on the condition that a unified procedure for the PC index calculations was defined. Unfortunately, this unification has not been accomplished yet, although a close cooperation between the Arctic and Antarctic Research Institute (AARI) responsible for the PCS index, and the Danish Meteorological Institute (DMI) formerly responsible for the PCN index, has solved most of the initial disagreements (Troshichev et al., 2006).

The relative importance of the solar wind dynamical pressure and the interplanetary geo-effective electric field for the various disturbance processes like the build-up of the current systems involved in magnetic storms and substorms is a subject of great controversy. Here, we shall specifically discuss the relations of the transpolar currents characterized by the PC index to the "external" solar wind parameters, E_M and P_{SW}, and we shall also consider the influence of "internal" auroral substorm processes parameterized by the auroral electrojet indices, primarily the AE and AL indices.

Following the introduction in section 1, section 2 briefly outlines the high-latitude response to solar wind forcing while section 3 more specifically discusses the response in the magnetic recordings to IMF variations. Section 4 outlines the calculation procedures for the PC indices, while section 5 discusses in more detail the relations between the PC indices and the merging electric field in terms of timing and amplitude response as well as the effects of the solar wind dynamic pressure on the PC index. Section 6 discusses the relations between the PC index and various further geophysical disturbances such as auroral electrojet activity, Joule and particle heating of the upper atmosphere, and the ring current activity. Finally, there is a concluding summary section.

2. High-latitude response to solar wind-magnetosphere interactions

It was early recognized (e.g., Spreiter et al., 1966) that the solar wind dynamical pressure to a large extent controls the general morphology of the magnetosphere, among other, its size and shape and also affects the internal magnetic field configuration. An important parameters to characterize the impact of the solar wind on the magnetosphere is the dynamical pressure, P_{DYN} :

$$P_{DYN} = D_{SW} \, V_{SW}^2 = M_{SW} \, N_{SW} \, V_{SW}^2 \tag{1}$$

where V_{SW} is the solar wind velocity while the density, D_{SW} , is a combination of solar wind number density, N_{SW}, and average particle mass, M_{SW},. After transition through the bow shock into the magnetosheath the solar wind dynamic pressure (P_{DYN}) is partly converted into thermal pressure (N_{SW} kT). At the magnetopause in Spreiter's model the thermal pressure from the solar wind in the magnetosheath is balanced by the internal "magnetic pressure" (i.e., N_{SW} kT=$B_M^2/2\mu_o$). The magnetic field generated by the boundary currents must cancel the geomagnetic field outside the magnetopause and adds to doubling the field strength just inside the boundary.

However, the solar wind electric field is in most cases the dominant factor for the high-latitude magnetospheric electric field structures and related plasma convection processes. A useful parameter is the "merging" (or "reconnection" or "geo-effective") electric field defined by Sonnerup (1974) and Kan and Lee (1979):

$$E_M = V_{SW} \, B_T \sin^2(\theta / 2) \tag{2}$$

This parameter is a combination of the solar wind velocity, V_{SW}, and the transverse component, B_T, ($B_T = (B_Y^2 + B_Z^2)^{1/2}$) of the interplanetary magnetic field (IMF) in the solar wind and includes a strong dependence on the field direction represented by the polar angle θ of the transverse component of the IMF with respect to the direction of the Z-axis in a "Geocentric Solar Magnetospheric" (GSM) coordinate system (i.e., $\tan(\theta) = |B_Y| / B_Z$, $0 \leq \theta \leq \pi$).

The merging electric field is the optimum parameter to characterize the energy transfer from the solar wind to the magnetosphere (Akasofu, 1979). It was shown theoretically by Kan and Lee (1979) that the time dependent power, P(t), delivered by the solar wind dynamo to the Earth's magnetosphere can be expressed in terms of the merging electric field, E_M, (defined in Eq.2) by:

$$P(t) = E_M^2 \, \ell_o^2 / R \tag{3}$$

where R is the total equivalent resistance connected to the solar wind-magnetosphere dynamo. The resistance includes the dissipation processes in the reconnection region, the energy used in driving the internal magnetospheric convection and building the ring current as well as the energy dissipated in the upper atmosphere by Joule heating and particle precipitation particularly in the auroral regions. The parameter ℓ_o is an interaction length ($\approx 7 \, R_E$), which, when projected along geomagnetic field lines to the polar cap ionosphere, defines the width of the open polar cap. Equation (3) indicates the importance of E_M to characterize the energy input although the concept is simplified and the parameters ℓ_o and R may not be quite well-defined or constant.

The interaction between the solar wind and the Earth's magnetosphere is sketched in Fig. 1 (from GEM Source book, 1999). The magnetosphere is delimited by magnetopause currents to provide the transition from the terrestrial magnetic field to the solar wind. In the diagram the direction of the IMF, depicted by the field lines, is assumed to be southward. In consequence, the IMF and the magnetospheric field merge at the front of the magnetosphere, where the terrestrial field is northward. The combined field reach to the ionosphere in the polar cusp region and is then dragged tailward over the polar cap and into the tail region by the plasma motion. In the middle of the tail region the neutral sheet of transverse currents separate the sunward magnetic field in the upper part of the plasma sheet from the antisunward magnetic field in the lower part of the plasma sheet.

The plasma flow just inside the magnetopause and in the northern and southern lobes at high latitudes is antisunward as in the solar wind, while the plasma motion is generally sunward at lower latitudes in the plasma sheet further inside the magnetosphere. The plasma velocity and pressure gradients close to the flanks of the magnetosphere generate the so-called "Region 1" (R1) field-aligned currents (FACs), downward to the ionosphere at the morning side, and upward from the ionosphere at the evening side. Part of the R1 currents that enter or leave the ionosphere closes across the polar cap while another part closes across the auroral oval to the "Region 2" (R2) field-aligned currents. The R2 FAC, upward at the morning side and downward at the evening side, connect to the ring current region at 4-6 Re and close through a partial ring current flowing from the morning to the evening side near equator (Iijima & Potemra, 1976a,b).

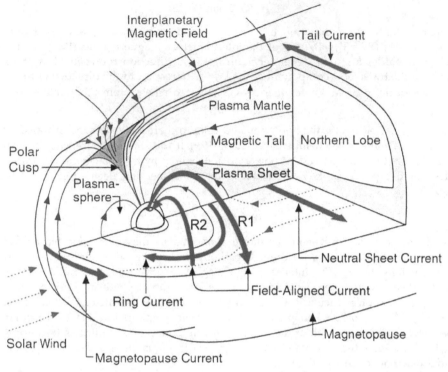

Fig. 1. Simplified sketch of magnetospheric structure (from GEM Source Book, 1999)

The polar ionospheric convection and current systems that could affect the magnetic variations measured from ground within the central polar cap could be divided into distinct categories, DP1, DP2, DP3 (or NBZ), and DP4 (or DPY) characterized by their relation to substorm activity and interplanetary magnetic field conditions (e.g., Wilhjelm et al., 1972; Friis-Christensen & Wilhjelm, 1975; Troshichev et al., 1986; Weimer et al, 2010). Each type of convection pattern comprise field-aligned currents (FAC) and horizontal Hall and Pedersen currents. The magnetic variations at ground can be assigned to a system of "equivalent currents", which in many cases will correspond to the ionospheric Hall currents that dominate the response at ground level, since the effects of the field-aligned currents and the horizontal Pedersen currents tend to cancel each other (Fukishima, 1969). The convective motion and the Hall currents are both transverse to the electric field but in opposite directions. Thus the convection patterns and the equivalent current patterns, although oppositely directed, will bear close resemblance. Both patterns are also close in shape to the equipotential contours.

The DP2 two-cell convection system comprising antisunward convection of plasma over the polar cap at all altitudes from the ionosphere up through the polar magnetosphere to the solar wind is the prevailing system except during intervals of strong northward IMF. However, the DP2 system may be modified to a greater or lesser extent by effects related to the DP1 (substorm) or/and the DP4 (DPY) convection systems. During northward IMF one or both vortices of the DP3 (NBZ) convection system could be formed at very high latitudes inside a weakened DP2 system prevailing at lower latitudes.

3. IMF and solar wind impact on polar geomagnetic variations

The main part, around 97%, of the geomagnetic field is generated by internal sources in the Earth's core and crust. The field from the core, the main field, has relatively slow (secular) variations in magnitude as well as in direction. The crustal remnant magnetization is constant on geological time scales. The remaining field contributions generated by external current systems and their induced counterparts in the ground are generally found by subtracting a set of base line values from the measured field components.

The external contributions can be subdivided into two fractions. One is associated with the effects of the slowly varying solar UV flux and the rather steady flow of solar wind plasma past the Earths magnetosphere during intervals with quiescent conditions. When this part, the so-called "Quiet Day" (QDC) variation, is also subtracted from the observed geomagnetic field, then the remaining field contributions are mainly related to the combination of enhanced solar wind velocities and an appreciable magnetic field in the solar wind, which may strongly influence the interaction of the solar wind flow with the Earths magnetosphere. The interaction manifests itself, among other, through enhanced ionospheric currents and associated magnetic variations in the Polar Regions.

Hourly values of the interplanetary magnetic field at the Earth's position derived from the ACE data base for the year, 2002, are shown in the top field of Fig. 2. These magnetic data are resolved in Geocentric Solar Magnetospheric (GSM) coordinates, B_X (toward the Sun), B_Y (perpendicular to the Earth-Sun line and to the Earth's magnetic dipole axis) and B_Z (completing an orthogonal system) almost in the direction of the (negative) dipole axis. Throughout the diagrams in Fig. 2 the heavy red continuous curves depict smoothed averages.

Typical examples of polar geomagnetic observations are also presented in Fig. 2. The middle field displays the data from Thule, Greenland, located at corrected geomagnetic (CGM) latitude of 85.30°. The bottom field displays data from Vostok, Antarctica, located at corrected geomagnetic latitude of -83.58°. The thin blue traces in the diagram show hourly average values of the field resolved in the H-component (geomagnetic North), E-component (East) and Z-component (down for Thule, up for Vostok) through the year 2002.

In these data several features are noteworthy:

i.   There are daily variations in all components during quiet as well as disturbed conditions judging the disturbance level from the IMF data in the top field. The main part of these variations is related to the rotation of the station beneath the more or less stationary polar cap current patterns.

ii. The polar magnetic data, particularly the E-component of the Thule observations, display small but significant steady changes through the year. These changes result from secular variations in the main field.

iii. There are low-frequency modulations on time scales from one to a few weeks seen in the interplanetary B_X and B_Y magnetic field components as well as in the ground-based Thule and Vostok data, most distinctly in the Z-component. These modulations are indicative of the interplanetary sector structure where the preferred field direction is either toward or away from the Sun. The associated changes in the polar magnetic field represent the so-called Svalgaard-Mansurov effect (Svalgaard, 1968; Mansurov, 1969).

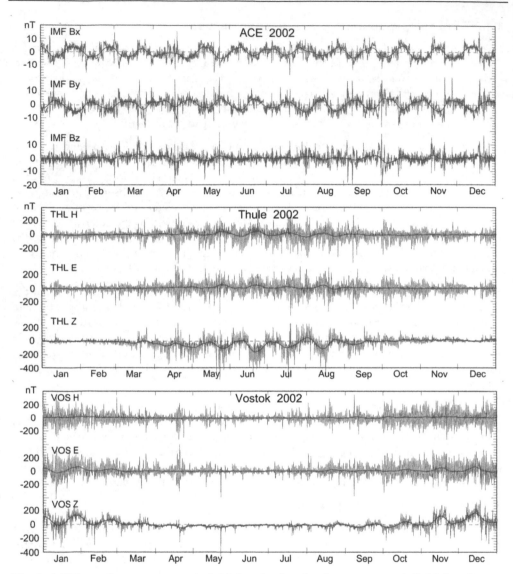

Fig. 2. ACE interplanetary magnetic field data (top) and polar geomagnetic field data from observatories in Thule, Greenland (middle), Vostok, Antarctica (bottom). Hourly values 2001.

iv. Strong high-frequency excursions, here on time scales of one to a few hours, are seen in the interplanetary data as well as in all three components of the ground-based data. In the ground-based data for Thule these excursions are strongest during the northern summer months occurring at the middle of the diagram. In the data from Vostok, correspondingly, the strongest excursions are seen in the southern summer months occurring at either ends of the diagram. The excursions on time scales from minutes to hours observed in the solar wind and at ground in the polar cap are strongly correlated. This correlation is the fundamental basis for defining a Polar Cap (PC) index.

Basically, the Polar Cap (PC) index is defined to be a proxy for the merging electric field by assuming a DP2 (or DP2-like) convection system with antisunward plasma convection and uniform sunward Hall current across the central polar cap. It is further assumed that there is proportionality between the horizontal magnetic field variations, ΔF, in the central polar cap and the merging electric field, E_M, and that the magnetic variations related to E_M have a preferred direction with respect to the Sun-Earth direction. Thus, on a statistical basis from observations of ΔF and E_M:

$$\Delta F_{PROJ} = S \cdot E_M + \Delta F_I \qquad (4)$$

where ΔF_{PROJ} is the projection of the magnetic disturbance vector, ΔF, to the "optimum direction", that is, the direction most sensitive to the merging electric field. The parameter S (often named α) is the "Slope", and the residual, ΔF_L (often named β) is the "Intercept" parameter, named from a visualized graphical display of ΔF_{PROJ} against E_M. Throughout the calculations we express magnetic fields in nT and the electric fields in mV/m.

In order to calibrate magnetic variations with respect to the merging (geo-effective) electric field we now define the PC index by the inverse relation:

$$PC = (\Delta F_{PROJ} - \Delta F_I)/S \quad (\sim E_M \text{ [in mV/m] }) \qquad (5)$$

The scaling parameters, that is, the optimum direction angle, the slope and the intercept, are found from Eq. 4 by regression analysis of polar observatory geomagnetic observations against interplanetary satellite data to include daily as well as seasonal variations. This scaling serves to make the resultant PC index independent of the regular daily and seasonal variations in parameters like the ionizing solar UV flux, that are not related to the merging electric field. Furthermore, the index is also now independent of the geographical position as long as the observatory is located within the central polar cap, where the magnetic variations are mostly associated with the rather uniform transpolar electrical currents that, in turn, relate mainly to the merging electric field in the solar wind and not to auroral substorm processes.

4. Calculation of PC indices

The calculation of PC indices according to Eqs. (4) and (5) requires a number of steps. Firstly, the calibration constants, optimum angle, slope and intercept, must be determined from Eq. (4) using lengthy series of correlated values of the merging electric field (Eq. 2) existing in the solar wind but adjusted in timing to apply to the Polar Cap, and geomagnetic variations corrected for secular variations and quiet day variations (QDC). Then the actual magnetic variations, again corrected for secular and quiet day variations, could be used in Eq. (5) with the calibration parameters, selected from tables spanning time of day and day of year, to provide PC index values (Troshichev et al., 2006; Stauning et al., 2006).

4.1 Satellite observations of solar wind and IMF parameters

Data for the merging electric field in the solar wind are needed for the derivation of the scaling parameters. During the time interval from 1975 to present, where high-quality digital magnetic data are available from Thule there are four main sources of solar wind plasma and interplanetary magnetic field data. These sources are IMP 8, Wind, Geotail, and ACE satellite data. Here we concentrate on IMP8 and ACE data for calculations of the scaling parameters, but Geotail and Wind satellite data can be used in special correlation studies since they offer further ranges of satellite positions with respect to the magnetosphere.

The IMP 8 satellite was launched 1973-10-26 into an orbit with inclination varying between 0 deg and 55 deg. Apogee and perigee were around 40 and 25 earth radii, respectively. Data are available up to June 2000. The spacecraft was located in the solar wind for 7 to 8 days of every 12.5 days orbit. Telemetry coverage has been varying between around 60 and 90%.

The ACE satellite was launched 1997-08-25 into a halo orbit about the L1 point. The satellite instruments and telemetry have provided almost 100% recovery of magnetic data and solar wind velocity data since the beginning of 1998. The satellite is still active. The main advantage of using ACE data is their regularity over the years, which are now spanning more than one full solar activity cycle. The main disadvantage of using ACE data is the large distance of around 240 Re (~1.500.000 km) from the Earth to the satellite. This large distance causes a delay of typically around 1 hour for the solar wind with its embedded magnetic field to travel from the satellite position to the encounter with the Earth's magnetosphere.

In most previous satellite-ground correlation studies IMP 8/ACE data have been referred to a fixed position at 12 Re in front of the Earth in order to provide a uniform basis. This modification is accomplished by shifting the data in time with an amount corresponding to the solar wind travel time from the actual satellite position to the GSE (=GSM) X-coordinate of the reference position assuming uniform conditions in planes perpendicular to the Sun-Earth line. For the recent analyses (Weimer & King, 2008), the satellite data have been shifted to the position of the bow shock nose (BSN). A detailed description of the shift is provided at the web site http://omniweb.gsfc.nasa.gov/.

For IMP 8 the upper limit on the X-coordinate is around 40 Re. This leaves a span for the solar wind travel from the satellite to the reference location to deal with distances between -12 Re (at $X_{GSE} = 0$) and +30 Re, which with typical solar wind velocities of 350 km/sec corresponds to delays ranging between around -4 min and +10 min. For ACE the typical delay is around 60 min between solar wind observations at the satellite and the encounter of the same solar wind volume with the Earth's magnetosphere (12 Re or BSN reference position). The above-mentioned delays are just typical values. In actual cases the precise satellite position and the observed solar wind velocity were used to calculate the relevant delays. With the above outlined procedure, all the relevant solar wind field and velocity parameters have been converted into 1-min samples at the reference position. For some of the statistics we proceed to calculate 5-min sorted averages (excluding max and min values, average of rest).

From the reference position (12 Re or BSN) it takes the E_M effects some time to propagate and affect processes in the central polar cap ionosphere. Precisely how the effects propagate to the polar cap ionosphere is still an open question involving magnetospheric convection processes, pressure gradients, field-aligned currents and coupling to the ionospheric potential and current systems. It is shown below that this time interval is around 20 min. Accordingly, the series of merging electric field values referenced to 12 Re or the bow shock nose should be further time-shifted by 20 min before being correlated with polar cap magnetic variations. In the further derivation of PC index parameters and other uses of IMF and solar wind data we have selected to base all calculations on IMP 8 and ACE data discarding the Wind and Geotail measurements except for specialized applications.

4.2 Handling of geomagnetic data

It is clear from Fig. 2, particularly in the trace for the geomagnetic E-component measured at Thule, that the magnetic field related to internal sources changes slowly. For Thule these secular changes in the components amount to several tens of nT each year. In Vostok the secular changes are smaller. These changes are unrelated to solar wind and ionospheric effects and must be corrected for before PC index values are derived. Thus the baseline caused by the "internal field" contribution should be derived. During midwinter, at night hours and for very quiet conditions, the "external field" contributions are at minimum, at most a few nT. Consequently, the winter night magnetic recordings are carefully monitored in order to detect the magnetic component values during extremely quiescent conditions. For Thule, these values, considered to represent the baseline related to the "internal field" contributions, are referenced to 1. January and kept since 1973 in a yearly updated "Quiet Winter Night Level", QWNL, table. It is assumed that the baselines for the individual components, with the accuracy needed for PC calculations, vary linearly through the year such that they could be interpolated from the values at day 1 one year to day 1 the following year, or extrapolated beyond 1 January through the actual year.

The "Solar Rotation Weighted" (SRW) calculations of the "Quiet Day Curve" (QDC) at DMI are explained in Stauning (2011). The method builds on superposition of a selection of quiet segments to build a QDC for each component separately for any given day. The samples are weighted according to the variability in the horizontal component. In the superposition of quiet recordings from different days, further weight functions are applied to give preference to intervals close to the QDC day in question and to intervals where the same face of the sun is directed towards the Earth. The solar rotation weight factor takes into account the regular variations in the solar UV radiation, in the solar wind velocity and in the IMF sector structure. The procedure is fully automatic and a quality parameter (the sum of weights) is provided for each hourly QDC value to enable a warning for poorly defined QDC values. Quadratic interpolations are used to provide QDC values with finer than one hour time resolution (e.g., 1-min samples). The baseline values and the QDC values calculated for each day of the data range are now subtracted from the recorded data to provide the series of magnetic variations to be used for PC index calculations. Composite plots of baseline-corrected magnetic recordings from January and July 2002 and the QDC values for these intervals are shown in Fig. 3.

4.3 Optimum direction angle

Searching for a proxy based on polar magnetic disturbances to represent the solar wind merging electric field (E_M = MEF = V_{SW} B_T $sin^2(\theta/2)$), we may increase the correlation between the horizontal disturbance vector ΔF (corrected for the quiet daily variations) and the E_M by projecting ΔF to a specific direction ("optimum direction"). In principle, this direction is perpendicular to the transpolar DP2 current. The direction is not entirely fixed in space but varies slowly with local time and season. The rotating horizontal magnetic vector is resolved in an X-component (northward in a geographical coordinate system) and a Y-component (eastward). The vertical Z-component is downward in the northern polar cap.

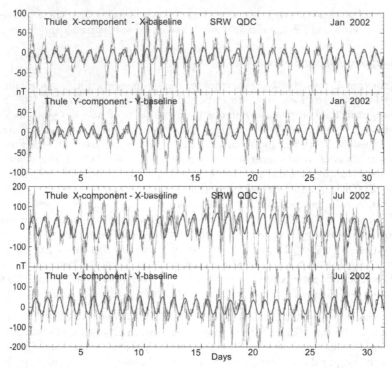

Fig. 3. Thule baseline-corrected magnetic data and QDC for X- and Y-components for January and July 2002.

For the horizontal components we first subtract from the raw data the baseline values and then the QDC values, i.e.:

$$\Delta X = (X_{RAW} - X_{BL}) - X_{QDC} \tag{6a}$$

$$\Delta Y = (Y_{RAW} - Y_{BL}) - Y_{QDC} \tag{6b}$$

Here, X_{BL} is the baseline value for the X-component defined yearly (every 1 January) and now adjusted to the actual day-of-year. X_{QDC} is the reference QDC level provided as tables of hourly values for each day and now adjusted to the proper time. The Y-component is handled correspondingly.

Using the + sign for the southern hemisphere and the – sign for the northern, the projection of the disturbance vector ΔF to the optimum direction is given through:

$$\Delta F_{PROJ} = \Delta X \cdot \sin (V_{PROJ}) \pm \Delta Y \cdot \cos(V_{PROJ}) \tag{7}$$

where the angle V_{PROJ} is defined as a function of UThr (UT time in hours) through:

$$V_{PROJ} = \text{Longitude} + \text{UThr} \cdot 15^\circ + \text{optimum direction angle (ODA)} \tag{8}$$

With this definition, the "Optimum Direction Angle" (ODA) is the angle between the dawn-dusk direction and the optimum direction for the magnetic variation vector. This angle is also the angle between the midnight-noon direction and the transpolar equivalent current direction, which by definition is perpendicular to the related magnetic disturbance vector.

The optimum direction angle values are calculated from statistical analyses to find the maximum correlation between the geomagnetic variations measured in the polar cap and the solar wind merging electric field values derived from interplanetary spacecraft data, which have been time-shifted to apply to polar cap effects. The direction depends on observatory location and varies with day of year, and UT time of day. With this procedure we have calculated and tabulated the optimum direction angle for each UT hour of the day and each month of the year.

4.4 Calculations of slope and intercept

It is the general assumption that there is a linear relation between the projected polar magnetic variations, ΔF_{PROJ}, and the solar wind electric field, E_M, and that this relation can be inverted and used to define a PC index by equivalence (cf. Eqs. 4 and 5). This relation assumes a DP2-type two-cell polar ionospheric "forward" convection, which is the most common convection mode. It applies to solar wind conditions where the interplanetary magnetic field (IMF) is either southward oriented or only weak in magnitude when northward directed. During conditions of strong northward oriented IMF a DP3 reverse convection system may develop in which the transpolar flow is sunward, while the return flow is antisunward.

With an overhead reverse current direction the magnetic deflections at ground are opposite of those of the forward convection mode. Accordingly, the projected disturbance vector may become less than the QDC level or even strongly negative. PC index values calculated during such conditions may turn out to be negative. Since the interplanetary merging electric field (E_M) by definition (Eq. 2) is always non-negative, then the concept of the PC index as a proxy for the E_M breaks down. Hence, reverse convection cases must be excluded from the calculations of slope and intercept.

In a least squares fit to estimate the regression of ΔF_{PROJ} on E_M from a comprehensive and representative data base, the deviations of the magnetic variations from the regression line are minimized whereby the slope, S, and the intercept parameter, F_I, are derived. With this procedure we have calculated and tabulated values of the slope, S, and the intercept parameter, ΔF_I, for each UT hour of the day and each month of the year.

4.5 PCN, PCS and combined PC index, PCC

The PCN index for the northern polar cap is based on data from the Danish geomagnetic observatory in Thule (Qaanaaq) in Greenland while the PCS index for the southern polar cap is based on data from the Russian geomagnetic observatory in Vostok at Antarctica. Parameters for the two observatories are listed in Table 1.

Station name	IAGA code	Geocentric Latitude	Geocentric longitude	Corr. geomag. latitude	Corr. geomag. longitude	UT at MLT noon
Qaanaaq	THL	77.47	290.77	85.30	31.11	15:04
Vostok	VOS	-78.46	106.87	-83.58	54.77	13: 02

Table 1. PC geomagnetic observatories in the northern and southern polar caps. Epoch 2000

Both the PCN and the PCS indices have been calibrated with respect to the merging electric field (MEF=E_M) in a statistical sense. Accordingly, the two index series are also mutually equivalent in a statistical sense. However, differences may arrive as the result both of different conditions in the two polar caps (e.g., different solar illumination) and of different response to forcing from the solar wind as well as different response to substorms. Such differences have the potential for interesting studies (e.g., Lukianova et al., 2002).

Another possible application of the two PC index series uses the possibility to define "summer" or "winter" PC index series based alternating on either PCN or PCS depending on local season. Thereby characteristic seasonal variations may emerge more clearly than otherwise by using one or the other hemispherical index series.

A further possibility is the combination of the two index series into one (Stauning, 2007). In the most common two-cell forward convection cases the projected magnetic variations and the PC index values are positive. However, with overhead reverse convection flow during NBZ conditions the magnetic deflections are opposite to those of the forward convection mode. Accordingly, the projected disturbance vector may become less than the QDC level or even strongly negative. PC index values calculated during such conditions may turn out to be negative. Since the interplanetary merging electric field by definition is always non-negative (cf. Eq. 2) then the concept of the PC index as a proxy for E_M breaks down.

Such cases of reverse convection and large negative PC index values are, by far, most common at daytime in the summer season. In these cases the forward convection mode may still prevail at the local winter season in the opposite polar cap and the PC index values derived there may be small but still positive like E_M. A simple way to accomplish a combined PC index would be to calculate the plain average values of the PCN and PCS data series. However, the problem with large negative PC index values in one hemisphere speaks for constructing a combined PC index (PCC) from non-negative values only. Accordingly:

$$PCC = [(PCN \text{ if } >0 \text{ or else zero}) + (PCS \text{ if } >0 \text{ or else zero})]/2 \qquad (9)$$

This option was tested in Stauning (2007) who examined the correlation between E_M and PCN, PCS, and the new PCC values, respectively. From this work, Fig. 4 presents hourly average values of E_M, PCN, PCS and PCC index values through the year 2000. The top field displays E_M values calculated according to Eq.2 from ACE satellite measurements of the interplanetary magnetic field and the solar wind velocity. The parameters measured at the ACE satellite have been shifted to apply to polar cap effects. The second and third fields from the top present PCN and PCS values using identical procedures in the calculation of the two index series. The bottom field presents the PCC index values derived from using Eq. 9.

Fig. 4 indicates a close similarity between the enhancements in E_M and the positive enhancements in PCN and PCS. These enhancements are, of course, also seen in the combined PC index, PCC. Further, the occurrences of negative PCN and PCS index values, preferably during the local summer seasons in the two hemispheres (May-August for PCN, November-February for PCS) should be noted. These events have no proportional (i.e., strongly negative) counterpart in E_M values. The relevant merging electric field values are just small at these times. The new PCC index displays the positive enhancements corresponding to the larger values of E_M and, by its definition, no negative values even during NBZ conditions.

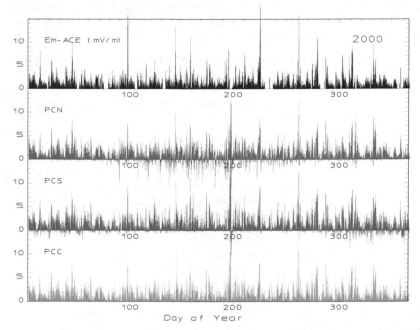

Fig. 4. Hourly averages calculated from 5 minute values of the merging electric field (top), Polar Cap PCN and PCS indices (middle) and new PCC index (bottom) through 2000.

5. Further relations of PC indices to solar wind parameters

The definition and the calculation of polar cap indices are based on an assumed linear dependence (cf. Eqs. 4 and 5) on the merging electric field, E_M (cf. Eq. 2). However, there are a number of additional effects, which could be of importance for various applications of the indices. Such effects, which to some extent violate the premises, include:

i.   Delayed PC index response to variations in the merging electric field.
ii. Non-linear effects such as saturation of PC index values at high levels.
iii. Response in the PC index to variations in the IMF Bx component not included in E_M.
iv. PC index response to solar wind dynamic pressure variations.

It is important to be aware of these modifications of the simple straight relations between E_M and the PC indices in order to recognize their signatures or consider whether they could deteriorate analyses of the relations between the PC indices and further terrestrial parameters.

5.1 Timing of PC index variations at changes in the merging electric field

When a change in the merging electric field, E_M, hits the magnetosphere then it takes some time for the polar cap convection pattern, and thus the PC index, to adjust to the new level. The timing issue could be analyzed in two ways. First, one could look for distinct variations, for instance, step-like changes in E_M, which could easily be recognized in the PC indices. Then the relative timing between Em and PC would provide the desired relation. Another approach would be to investigate statistically the correlation between E_M and PC index values with a variable delay in order to disclose the optimum value. We shall proceed both ways.

The top field of Fig. 5a displays a plot of superimposed traces of step-like varying E_M. The data have been shifted to a position of 12 Re in front of the magnetosphere (close to the average position of the bow shock nose). The individual traces have been positioned such that the positive step from a low to a high value occurs at relative timing T=0. The bottom field of Fig. 5a displays the corresponding variations in PCN. The heavy traces in the two plots mark the average variation.

It is easy to see the sudden step in the average E_M trace (top field) from close to zero up to around 2.5 mV/m. The average PCN index values in the bottom field display a gradual change from almost zero and up to 2.5 units, where it levels with the value of E_M. The change in PCN appears to start 9-10 min after the step in E_M. The average PCN index values reach half the final level at around 20 min after the E_M step, while it takes 30-40 min to reach the final level.

The corresponding variations at a negative step in E_M are displayed in Fig. 5b. In the upper field a number of step-like varying Em traces have been superposed with the step placed at relative timing T=0. From the average variations in the corresponding PCN index values it is seen that the change starts around 10 min later than the step in E_M. The average PCN value reaches half the level prevailing before the change at around 20 min after the negative E_M step while it takes around 50 min to reach the final low level.

In summary, following a positive or negative step-like change in the merging electric field impinging on the front of the magnetosphere it takes around 10 min before the change is sensed and the PC index starts to change. At around 20 min after the E_M step the PC index has reached half the final level, while it takes 40-50 min before the change is completed. The positive steps in E_M propagate a little faster to affect the PC index than the negative steps.

The analysis of corresponding features based on a larger and more general selection of data is presented in Fig. 6. For the epoch from 1990 to 1999, PCN data have been plotted against E_M in such diagrams with variable delays from 0 to 30 min in steps of 5 min. The IMP8 satellite data were selected for this study because of the proximity of the satellite to the magnetosphere whereby the uncertainty in the time-shift is minimized (compared, e.g., to ACE or Wind data). At each step the linear correlation coefficient, RX, was calculated along with the parameters S0 (average deviation), S1 (absolute deviation), and S2 (standard deviation). NX is the total no. of samples. The number of samples in each bin of 2 mV/m in E_M is indicated by the size of the large dots and the standard deviation is indicated by the bars. The example shown here uses a delay of 20 min. From the values of RX at different delays, the optimum delay was estimated to be 18 min. At this delay the correlation coefficient was found to be RX=0.748.

A different representation of the correlation between PCN and E_M is displayed in Table 2. Here, the PCN data and IMP8 E_M data through 1990 to 1999 are again correlated. The data have been divided in selections based on the values of the bow shock nose distance, BSN-X, supplied with the satellite data. In addition to confirming that the delay from bow shock nose to polar cap is around 20 min, the correlation study presents further details specified in Table 2.

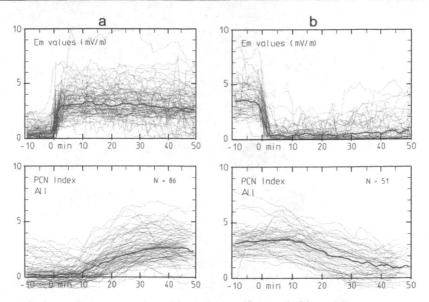

Fig. 5. PCN index variations at steps in the merging electric field at 12 Re.

BSN-X	8	9	10	11	12	13	14	15	16	Re
NX	397	2082	9401	31068	75318	87184	45806	12568	3136	
RXmax	0.45	0.59	0.71	0.74	0.70	0.69	0.69	0.70	0.72	
DTm	12.3	15.5	16.3	17.2	17.4	18.8	20.9	23.2	27.4	min
E_M av	7.7	4.6	2.7	1.7	1.2	0.9	0.7	0.7	0.7	mV/m
PCNav	6.3	4.2	2.4	1.7	1.2	0.8	0.6	0.6	0.6	

Table 2. Results from correlation of PCN with values of E_M at bow shock nose (BSN).

From Table 2 it is clear that for the majority of cases (90%) the BSN distances range between 11 and 15 Re, which is close to the X=12 Re used as a reference position in other analyses. It is also seen that the delay, generally, decreases with decreasing BSN distance. The decreases are larger than the reduction by 17 sec for each 1 Re decrement in distance, which is the time lapse for a typical solar wind speed of 400 km/s. The reduced correlation for the stronger cases (low BSN-X) could indicate that the strongly reduced travel time relies on the mixing of different propagation mechanisms, for instance, effects from accompanying solar wind dynamical pressure enhancements that propagate faster to the Earth than the E_M -associated effects.

Another remarkable feature is the close relation between E_M and PCN for the different ranges of BSN-X. It could be expected that the bow shock moves closer to the Earth when the solar wind is intense, that is, for cases of high solar wind speed and strong, southward IMF in which case the merging electric fields are enhanced. However, it is noteworthy that the average PCN index for each BSN-X interval corresponds closely to the average E_M value except for the highest levels where some saturation of PCN is evident.

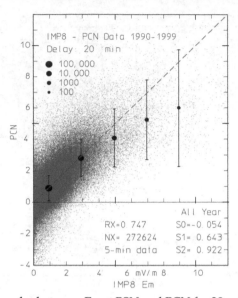

Fig. 6. Correlation scatter plot between E_M at BSN and PCN for 20 min delay.

5.2 PC saturation effects

At high levels of the solar wind forcing, saturation effects are found in many terrestrial parameters, for instance in the polar cap potentials, in the convection velocities, and the cross-polar cap dimension, or in the length of the reconnection line. Borovsky et al. (2009) list 9 different models to explain polar cap potential saturation.

A relevant question is now whether the polar cap indices just reflect the general saturation in the magnetospheric response to strong solar wind forcing or have saturation patterns different from other geophysical disturbance phenomena. Fig, 7 displays in a scatter plot the relations between the combined PCC index and the merging electric field. The data base has been constrained to intervals of 4 days through each of 82 geomagnetic storm intervals (Dst<-100) from the epoch 1995-2005 in order to focus on the strong events with large values of the PC indices. The PCC indices have been preferred over the hemispheric PCN or PCS indices since possible saturation effects are globally extended. From the scatter plot in Fig. 7 the saturation effect is evident. The dotted curve included for reference represents the simple saturation model:

$$PCC = E_M \Big/ \sqrt{\left(1+\left(E_M/E_0\right)^2\right)} \qquad (10)$$

The asymptotic electric field value used here is: E_0=10.5 mV/m. The fit to the actual average PCC index values is quite good. Hence, this saturation model could serve to provide a basis for comparison of saturation in the PC index data with saturation effects in further parameters. The model (Eq. 10) might even provide some guidance in a discussion of possible saturation mechanisms.

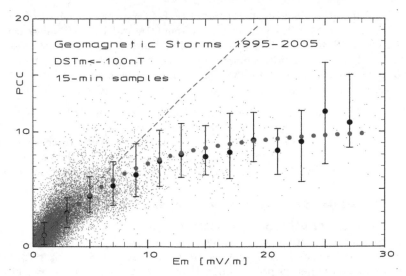

Fig. 7. PCC indices vs. related merging electric field values for all major magnetic storms (Dst<-100 nT) during 1995-2005.

The saturation effect in the PC index may also be seen as a limitation in the response to large values of the IMF Bz parameter. This is illustrated in Fig. 8. The figure displays in a scatter plot the ratio of PCN over E_M as a function of the IMF Bz component. The heavy dots mark the average values within intervals of 2 nT in the Bz component. The Bz component is already included in the expression for E_M (cf. Eq. 2). Ideally, the points should thus be positioned close to and symmetrically around the horizontal dashed line, which indicate a ratio of unity, and should not show any systematic dependence on the IMF Bz (or any other) component.

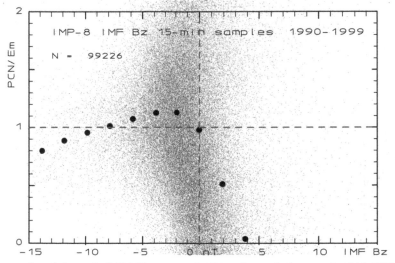

Fig. 8. Scatter plot of the ratio PCN/E_M versus IMF Bz through the epoch 1995-1999.

The left half of the plot displays the ratio PCN/E_M for negative values of the IMF Bz component (southward IMF). The ratio increases from 1.0 at IMF Bz ~ 0 to values between 1.00 and max. 1.13 for IMF Bz values between 0 and -8 nT. Starting from the max. level at an IMF Bz level at around -4 nT, the ratio systematically decreases in an almost linear trend to smaller values ending here at a ratio of 0.8 for IMF Bz ~ -15 nT. This systematic behaviour may – like the above saturation model – provide some guidance to the saturation mechanism.

The right part of Fig. 8 displays the ratio PCN/E_M for positive values of the IMF Bz component (northward IMF). From an average value of this ratio close to unity at IMF Bz ~ 0 the ratio rapidly decreases to zero at IMF Bz ~ +4 nT and thereafter reaches large negative values since the PCN values tend to become negative while the E_M values approach zero in consequence of the IMF direction getting close to due northward in some cases (cf. Eq.2).

5.3 Effects from the IMF Bx component on PC index values

The large-scale systematic variations in the IMF Bx component define the sector structure in the solar wind. The Bx component is positive when the IMF is pointing toward the Sun in "toward" sectors and negative in "away" sectors (Svalgaard, 1968, 1972, 1973; Mansurov, 1969). The IMF Bx component is not included in the expression for the merging electric field (Eq. 2). Hence the scaling of polar magnetic variations to construct a polar cap index is assumed to be independent of this component.

Fig. 9 displays the ratio of PCN over E_M value plotted against the IMF Bx value. The scatter plot has the same format as the plot displayed in Fig. 8. The heavy dots indicate average values of the ratio PCN/ E_M through intervals of 2 nT in the Bx component. These values are all within 0.9 to 1.1, that is, close to unity through most of the displayed IMF Bx range from - 15 to + 15 nT . Ideally, the average value of ratio PCN/ E_M should be close to unity regardless of the IMF Bx value. Thus, Fig. 9 provides a confirmation that the IMF Bx component has little influence on the PC values.

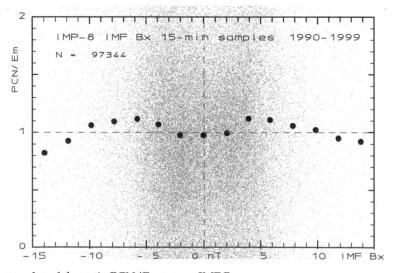

Fig. 9. Scatter plot of the ratio PCN/E_M versus IMF Bx

5.4 Effects from variation in the solar wind dynamic pressure on the PC index

The dynamic pressure in the solar wind (cf. Eq. 1) has profound influence on the general morphology of the magnetosphere. The pressure balance at the magnetopause is an essential factor for determining the dimensions of the magnetosphere (e.g., Spreiter et al., 1966; Shue et al., 1997) and for the location of the bow shock as well (e.g., Farris & Russell, 1994). Furthermore, the solar wind dynamic pressure strongly affects the width of the auroral oval causing auroral activity to expand equatorward during intervals of enhanced pressure (Newell & Meng, 1994). However, the solar wind dynamic pressure and its variations have relatively small effects on the polar cap index compared to the effects of the merging electric field and its variations.

The PC index variations with solar wind dynamic pressure could be divided into 3 types comprising effects of different steady levels, gradual changes, and fast (impulsive) variations, respectively. The sample duration and averaging procedures should be selected accordingly. The variations in the PC index level with solar wind dynamic pressure level are illustrated in Fig. 10. The diagram is based on 15-min average values of the combined PCC index, the solar wind merging electric field (E_M), and the dynamic pressure (P_{SW}) for a series of 82 selected magnetic storm intervals during 1995-2005. Each interval comprises 4 days of which the first includes the storm onset. This selection holds a fair amount of enhanced parameter values. Each of the small points displays the ratio of PCC over the E_M value (vertical axis) against the dynamic pressure (horizontal axis). The horizontal dashed line reflects the "ideal" unity ratio between PCC and E_M. The large "dots" indicate average values of the ratio of PCC over E_M for each interval of width equal to 2 nPa in dynamic pressure. The error bars represent the 68% "standard deviation" range for the distribution of points above and below the average values.

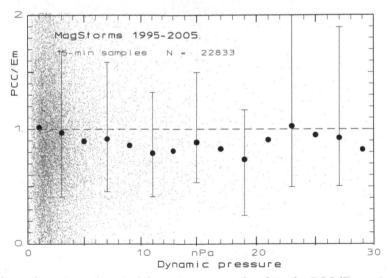

Fig. 10. Effects of varying solar wind dynamic pressure level on the PCC/Em ratio.

Like seen in other scatter plots, the scatter of points in Fig. 10 is quite large. However, it is readily seen that the average ratio of the PCC index over the merging electric field is very close to unity for the bulk of data points representing small and moderate levels of the dynamic pressure up to around 2 nPa. For levels of the dynamic pressure from 2 up to around 10 nPa the PCC index is less than the corresponding E_M value. For still larger values of P_{SW} above 10 nPa the ratio of PCC over E_M is again close to unity but the scatter increases due, among other, to the sparse statistics. Accordingly, through the wide range displayed in Fig. 10 the steady level of the solar wind dynamic pressure has little effect on the polar cap index values.

The relation of the PC index values to gradually changing solar wind dynamic pressure conditions is displayed in Figure 11. The solar wind pressure values used in the figure are 5-min averages of the slope of the pressure variations with time. The solar wind data have been time shifted to apply to the polar cap (cf. section 5.1). The selection of data used to determine the relation of PCC/E_M to the time derivative in the solar wind dynamic pressure is the same set of storm intervals as used previously. The PC index values used in Fig. 11 are 5 min averages of the combined PCC index.

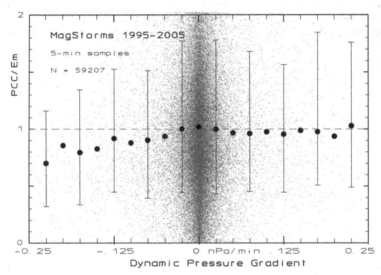

Fig. 11. Effects of dynamic pressure gradient on the ratio of PCC over E_M.

The large dots depict averages of all data samples within each interval of width equal to 0.025 nPa/min. For every other dot error bars have been drawn to indicate the 68% range for values above and below the average value. The average ratio of PCC to E_M (in mV/m) should be close to unity. This is clearly the case for the bulk of data points with time derivative values close to zero or positive. For large negative values of the time derivative the average ratio of PCC over E_M values is less than unity. The effects on the ratio PCC/E_M and thus on the PC index is fairly small within the range of gradient values displayed in Fig. 11.

The effects on the Polar Cap index of sudden (shock-like) pressure variations are depicted in Fig. 12. The top field displays the dynamical pressure variations referred to the reference position (12 Re). Each of the PCN index traces in the bottom field displays a sequence of an

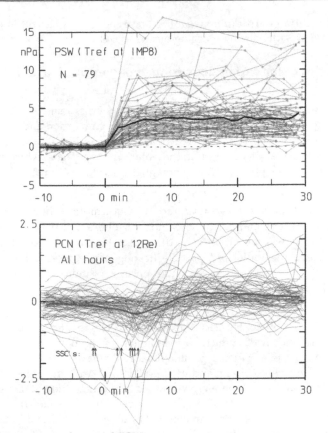

Fig. 12. PCN variations at solar wind PSW steps.

initial negative impulse followed by a positive enhancement in the index value (e.g., Lukianova, 2003; Huang, 2005; Stauning & Troshichev, 2008). The negative peak in the PC index occurs within 2-10 min following the pressure step referred to the reference position at 12 Re in front of the Earth. The width of the negative pulse is typically 3-5 min and the amplitude is around -0.5 to -1.0 units in the PC index. The positive pulse, typically, has a width of 10-20 min and amplitude of 0.5 to 1.0 units. The peak positive amplitude is typically reached ~15 - 20 min after the pressure step impacts at 12 Re. The timing implies that the pressure effects propagate from the reference position (12 Re) to the central polar cap in only a couple of minutes. This is much faster than the propagation of effects related to the E_M and thus suggests different propagations processes (e.g., Araki, 1994; Stauning and Troshichev, 2008).

In summary, the PC index variations with varying level of the solar wind dynamic pressure and with impulsive variations are usually limited to range between – 0.5 and + 1.0 units. The variations at impulsive pressure enhancements could in rare cases amount to a couple of units, which is still fairly modest compared to potential variations with the merging electric field (cf. Fig. 7). The response time for the initial negative pulse in the PC index is a few minutes while the positive trailing impulse develops in 10-15 min suggesting propagation modes different from the usual FAC and convection-based timing of variations in the PC index values with varying merging electric fields.

6. Relations of the polar cap indices to further geophysical disturbance phenomena

The solar wind flow past the magnetosphere generates a cross-magnetosphere electric field along the magnetopause. This field may feed large-scale convection and current systems in the magnetosphere, particularly during southward IMF conditions, and is partly mapped down to create the potential distribution in the polar cap ionosphere, which in turn generate the ionospheric currents responsible for the magnetic variations transformed into the polar cap indices.

Thus, the polar cap indices could be considered to be indicators of the power conveyed from the solar wind to the Earth's magnetosphere and ionosphere. The energy extracted from the solar wind is used to power, among others, the transpolar current and convection systems, the auroral electrojets, the Joule and particle heating of the thermosphere, and the ring currents. Hence, close relations between the polar cap indices and these disturbance phenomena could be expected.

Fig. 13 presents an example case of solar wind and geophysical parameters for a geomagnetic storm event on 12 Sept. 1999. The case has been selected to display data from the three satellites, ACE, Wind and IMP8, which were simultaneously in operation during the event. The ACE and Wind magnetic field and plasma data are displayed in the upper two fields. The traces are plotted in different colours and given different marks. In the next lower fields the merging electric fields and the dynamic pressure based on data from all three satellites have been plotted. The three lower fields display the auroral activity characterized by the AL and AE indices, the ring current intensity characterized by the SYM, ASY, and Dst indices, and the polar cap currents characterized by the PCN, PCS, and PCC indices. The average E_M (MEF) and Psw values have been plotted in the bottom field for reference.

The onset of activity at ground level is indicated by the arrow marked "SSC" (Storm Sudden Commencement) at around 04 UT. During the next ~ 6 hours the Earth was hit by a cloud of enhanced solar wind density causing a strong increase in the dynamical pressure (middle field). This increase, replotted in the bottom field, had no marked influence on the PC indices and also little influence on the auroral electrojet and the ring current indices. At 09 UT the IMF Bz turned southward and the merging electric field increases strongly. This time the PC indices increased markedly and subsequently the PC indices track E_M quite close, and the AE and AL traces indicate strong auroral activity. In the following sections the relations between the polar cap indices and further geophysical disturbances are discussed.

6.1 PC indices and polar cap potentials

The cross polar cap potential difference, Φ_{PC}, is an important parameter to characterize polar cap potentials. This parameter could be investigated by radars, for instance the SuperDarn backscatter radar system (Greenwald et al., 1995), or by direct measurement of the electric field from satellites traversing the polar cap (e.g., Reiff et al., 1981; Rich and Hairston, 1994; Boyle et al., 1997; Hairston et al., 1998).

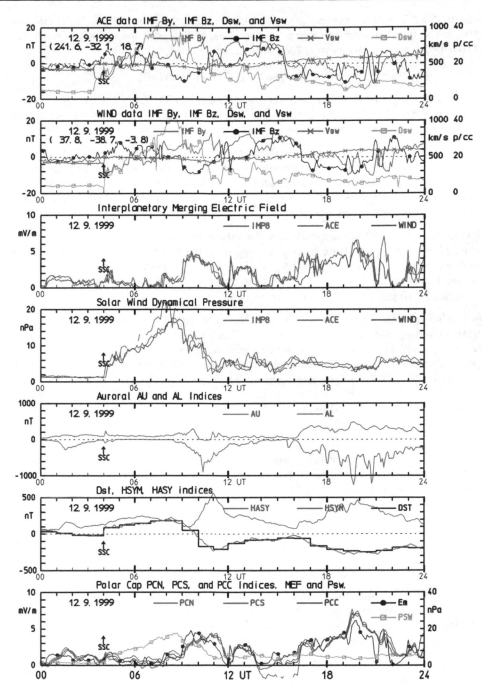

Fig. 13. Example plot of solar wind data from ACE, Wind, and IMP8, and indices for geophysical disturbances, viz. auroral electrojet, ring current, and polar cap currents.

Based on DMSP measurements, comprehensive studies by Boyle et al. (1997) suggest an empirical relation between solar wind parameters, B, V_{SW}, and the IMF clock angle θ, and the cross polar cap potential drop, Φ_{PC}, of the form:

$$\Phi_{PC} = 10^{-4} \, V_{SW}^2 + 11.7 \, B \, \sin^3(\theta/2) \, [kV] \tag{11}$$

Unfortunately, most of the observations relate to quiet or moderately disturbed conditions only. Thus, there were few measurements of Φ_{PC} beyond 100 kV. Hence, the expression does not provide the saturation effects observed during strongly disturbed conditions.

Further developments of empirical relations by Siscoe et al. (2002) provide relations where the height-integrated Pedersen ionospheric conductivities, assumed to be uniform across the polar cap, are included. A problem here is whether to use northern and southern hemisphere conductivities for northern and southern passes, respectively, or to use composite conductivities since the two polar caps are electrically coupled.

Hairston et al. (2005) studied specifically the saturation effects seen during strong magnetic storms. They observed that the cross polar cap potential rarely exceeded 200 kV and only in exceptional cases reached up to 250 kV. The "turn-over" from the non-saturable relation (Eq. 11) from Boyle et al. (1997) and the saturated range would occur at solar wind merging electric field (E_M) values between 4 and 8 mV/m.

This is also the range of E_M values where the polar cap indices start to show substantial saturation effects. Interestingly, the potential data presented in Fig. 10 of Hairston et al. (2005) could well be represented by an expression similar to the above Eq. 10. Their Fig. 10 displays the cross polar cap potential drop in a scatter plot (50 points) against the merging electric field in the solar wind ranging from 0 to 40 mV/m. On basis of these data and the PCC values depicted in Fig. 7 and Eq. 10 here, the relation between the polar cap index, PCC, and the potential would read:

$$\Phi_{PC} \approx 20 \, PCC + 15 \, [kV] \tag{12}$$

The relations between polar cap index and cross polar cap diameter and voltage was investigated by Troshichev et al. (1996). They conclude that the cross polar cap voltage and radius could be expressed by:

$$\Phi_{PC} \approx 19.35 \, PC + 8.78 \, [kV] \tag{13}$$

$$R(morning) = -.12 \, PC^2 + 2.5 \, PC + 11.0 \, [deg] \tag{14a}$$

$$R(evening) = -.20 \, PC^2 + 3.0 \, PC + 12.5 \, [deg] \tag{14b}$$

Their expression for Φ_{PC} (Eq. 13) is very close to the above expression (Eq. 12) deduced from the observations reported by Hairston et al. (2005). Their expressions shown here in Eqs. 14a,b for the polar cap radius (angular distance from the pole) show the non-linear enlarging of the open polar cap with increasing values of the PC index.

In the analyses by Troshichev et al. (2000) they derive a non-linear expression for the average ionospheric electric field in polar regions:

$$E_{PC} = 9.29 + 3.76 \, PC - 0.11 \, PC^2 \, [mV/m] \tag{15}$$

For the cross polar cap potential difference, the nonlinear quadratic reduction of the electric field strength is counteracted by the expansion of the polar cap radius. For a moderately strong event the polar cap index could typically have a magnitude PC = 5. In that case the radius R ≈ 21.5° and electric field E ≈ 25.3 mV/m would give a cross polar cap voltage Φ_{PC} ≈ 121 kV, not so different from the results from Eq. 12 (Φ_{PC} ≈ 115 kV) or Eq. 13 (Φ_{PC} ≈ 106 kV).

Based on SuperDARN radar measurements Fiori et al. (2009) derived a similar relation between Φ_{PC} and the PCN index:

$$\Phi_{PC} \approx 35 + 12 \text{ PCN [kV]} \tag{16}$$

Their values of the cross polar cap voltage are somewhat smaller than those derived from the above expressions, Eqs. 12 and 13. However, the linear (non-saturated) relation between the cross polar cap voltages and the PC indices is maintained.

In summary, the cross polar cap potentials are linearly related to the PC index values, notably to the PCC indices, which reflect the global situation more reliably than the hemispherical indices. The ionospheric electric fields in the polar cap display non-linear saturation effects also in relation to the polar cap index, but for the cross polar cap voltages these effects are to some degree counteracted by the expansion of the polar cap during strong events to result in a linear relation between Φ_{PC} and the PC index.

6.2 PC indices and the auroral electrojets

The Auroral Electrojet indices were introduced by Sugiura and Davis (1966) to measure the over-all electrojet activity in the auroral zone. Index values are based on magnetic data from a number of observatories, usually between 8 and 12, located in the auroral zone. AE index values are derived as the differences over the array of observatories between the upper envelope (AU) and the lower envelope (AL) for superposed variations in the horizontal (H) components. 1-min AU/AL/AE index values scaled in nT are derived and made available by WDC-C2 (Kyoto). AU index values are mostly related to the eastward electrojet in the post-noon sector. The AL index values are most often dominated by contributions from the westward auroral electrojet currents in the morning and midnight sectors. The westward electrojet currents are strongly intensified during substorm activity. Furthermore, substorms are effective in generating intense radiation of energetic auroral particles that, in turn, precipitate to produce increased upper atmospheric electron densities. The resulting increases in ionospheric conductivities in the auroral regions, generally, further enhance the electrojet currents.

Auroral activities (e.g., substorms) are linked to the effects of the solar wind on the magnetosphere. The primary parameter in the solar wind is the merging electric field, E_M, but the dynamic pressure, Psw, also plays a role mainly for the position and width of the electrojet regions. The polar cap magnetic activity characterized by the PC indices could be considered representative of the input power from the solar wind to the magnetospheric processes driving auroral activities. Thus the auroral and polar cap activity levels must be related.

The correlation coefficients between the PCN or PCS indices and the AE index were calculated by Troshichev et al. (2007) for each UT hour through 1998 to 2001. An example of the results is displayed in Fig. 14. The correlation is best at local winter where it would range between 0.80 and 0.85. During local summer the correlation is lower and ranges between 0.7 and 0.75. The results displayed in Fig. 14 also illustrate the solar cycle dependency. The correlation coefficients for PCN and PCS against AE are generally higher by 0.05-0.10 during solar minimum year (1998) than during solar maximum years (1999-2001).

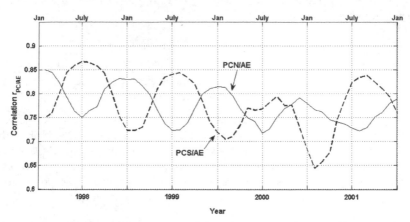

Fig. 14. Seasonal and solar cycle variations of the AE – PC correlations. (from Troshichev et al., 2007)

Another example from a recent correlation study is presented in Fig. 15. Here, values of the auroral electrojet index, AE, are plotted against corresponding values of the combined polar cap index, PCC, using a delay of 5 min. (PCC changes first). Note, that the correlation coefficient, R=0.816, for the full year is considerably higher than the above coefficients based on the individual PCN or PCS indices. Also note that the average AE indices show little indication of saturation (but larger spread) up to PCC levels of ~ 9.

The timing between sudden changes in the solar wind conditions and related changes in the polar cap and, for instance, the onset of auroral substorms as indicated by a sudden negative excursion in the AL index or a sudden increase in the AE index, was analyzed by Janzhura et al. (2007). They investigated isolated substorm events where the growth phase and the onset could be distinctly identified and related to the timing and magnitude of an increase in the PC indices caused by a gradual increase in the solar wind electric field, E_M. They found that when the E_M amplitude (and the PC index) remained below a level of 2 mV/m, then there was no clear indication of directly related substorm onsets. For E_M rising to amplitudes between 2 and 6 mV/m there would be a substorm onset following a growth phase where the PC and AE indices would gradually increase. The average growth phase duration varied systematically with the E_M amplitude from 60 min down to 0. For E_M amplitudes (and related PC levels) above 6 mV/m the substorm activity would immediately follow the E_M increase without any significant delay.

A relevant question is now whether the general activity level of auroral processes as such may influence the polar cap activity level above the direct effects of the solar wind parameters, notably the merging electric field, E_M. Figure 16 presents the relations between

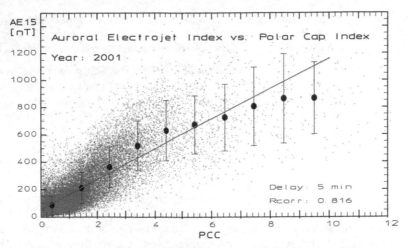

Fig. 15. Scatter plot between values of auroral electrojet indices, AE, and polar cap indices, PCC, through 2001.

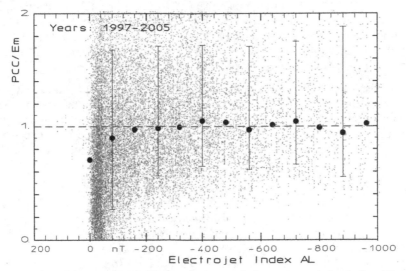

Fig. 16. Scatter plot of the ratio PCC/E_M versus auroral electrojet activity index AL

the ratio of PCC over E_M and the level of the (usually negative) auroral electrojet index, AL. The horizontal dashed line represents the ideal unity ratio between PCC and E_M. A noticeable feature is the variation from a relatively low value of PCC/E_M at around 0.7 for AL values close to zero to almost unity for AL equal to -150 nT. This range comprises quiet and weakly disturbed conditions. For magnetically disturbed conditions with still larger negative values of AL beyond -150 nT, there is hardly any variation in the average ratio of PCC over E_M. There is a substantial scatter in the distribution of values but the average ratio stays close to unity through the range from -150 up to -1000 nT in the auroral electrojet index, AL. Beyond that range the statistics is too sparse for conclusions. The variations in the ratio PCC/E_M at weak to moderate auroral activities (AL>-150 nT), no doubt, relate to the increase in conductivities through ionization by substorm-associated particle precipitation to reach a level adequate to support the ionospheric currents responsible for the PC indices.

A different question is whether the individual steps of the substorm cycle, comprising growth phase, onset, expansion phase, and recovery, affect the polar cap activity variations around the mean level predominantly defined by the solar wind electric field, E_M. In order to investigate this question we have "synchronized" PC index variations to the onset of substorms. The results are displayed in Fig. 17. The data refer to cases with steady solar wind conditions from at least 1 hour before the onset time until 2 hours after. Epoch time zero in the fields is the substorm onset time as defined by a sudden negative excursion in the AL index.

The top field displays the solar wind electric field, E_M, (referred to 12 Re reference position) through an interval spanning from 1 hour before to 3 hours after onset. On top of the individual traces the heavy line marks the average value. The next lower field displays the values of the auroral electrojet indices AU (upper traces) and AL (lower traces). The heavy lines mark the average values within each of the two index data sets. The sharp downward change in AL at epoch time zero is readily identified. The negative peak value in AL is reached at around 10-15 min after onset. The AU indices, notably their average values, hardly indicate any change at all. The next lower field displays the PCC index variation on the same time scale as the AU/AL display. A remarkable feature here is the peak at around 30-40 min after substorm onset, that is, at around 20-25 min after the negative peak in AL (or the corresponding positive peak in AE=AU-AL).

The bottom field repeats the average E_M variations (red line) and average PCC variations (blue line) while the heavy line displays their ratio, PCC/E_M, using the scale to the right. It should be noted that the PCC/E_M ratio has a marked minimum value of approximately 0.7 at around epoch time -30 to -20 min and a maximum value of around 1.3 at epoch time 35-40 min after onset. At 120 min there is another minimum at a value close to 0.8. Thereafter the PCC/E_M ratio approaches unity.

Using substorm terminology for the PCC/E_M variations, there is clearly a "pre-onset" or "build-up" phase with minimum PCC/E_M ratio prior to substorm onset. There is no steep onset but a gradual "expansion" phase with increasing ratio leading up to a peak value at 35-40 min after substorm onset. Then follows a "recovery" phase with gradual decrease in the PCC/E_M ratio. In the recovery phase the ratio might have reached the low pre-onset level (0.7) at around 120 min after onset had there not been some new events at this time to raise the average PCC/E_M level. These new events are spotted in the PCC plots (second field from bottom). Correspondingly, the pre-onset level is probably influenced to some degree by the increased level experienced during the recovery phase following the preceding event.

In summary of the above analyses the PC index provides a fair indicator of subsequent substorm activity. For quiet conditions with PC<~2 units there are hardly any substorm onsets. The auroral electrojet activity indices AE and AL track possible variations in the PC index. If the PC index increases from quiet to more disturbed conditions then substorm onsets are likely to occur as the PC index increases beyond ~2 units with a delay depending on the actual magnitude of the PC index, typically a few tens of minutes. If the PC index takes values above ~5 units then substorm activity follows immediately. There is some

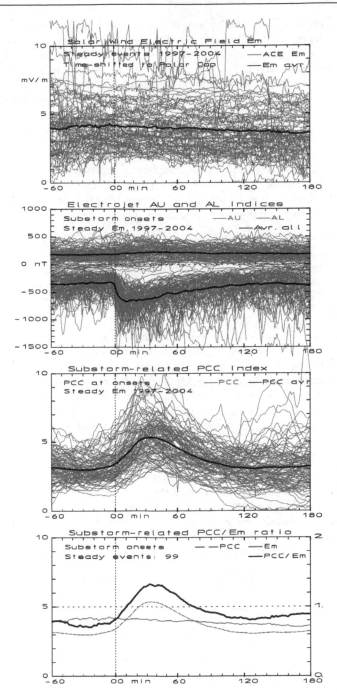

Fig. 17. Solar wind electric fields (top), AL and AU indices, PCC indices, and averages of Em and PCC and their ratio PCC/Em (bottom) for cases with steady interplanetary conditions.

coupling from auroral activity to the amplitude of the polar cap indices during weakly disturbed conditions probably related to the changes in ionospheric conductivities resulting from precipitation of substorm-associated auroral particles. The overall correlation between auroral activity represented by the AE indices and the polar cap indices, PCN or PCS, ranges between 0.60 and 0.85 depending on seasonal and solar cycle conditions. Using the combined PCC index enhances the correlation to reach an average of around 0.82.

6.3 Relations between PC indices and Joule and auroral particle heating

The thermosphere, particularly in the auroral regions, is heated by the amount of energy deposited by Joule heating and by precipitating particles. The heating generates thermospheric vertical motions and composition changes. The associated expansion of the atmosphere could have grave consequences for the tracking of satellites and for other spacecraft operations through the varying drag on space objects.

In various models to calculate drag coefficients, parameters such as the F10.7 cm flux (proxy for solar EUV radiation), the planetary Kp index, or the auroral electrojet index have been used to derive Joule heating values (e.g., Tobiska et al., 2008). Further attempts to calculate Joule heating include estimates of Joule heating in response to IMF input based on large-scale modelling of electrodynamic parameters (e.g., Chun et al., 1999, 2002; Weimer, 2005; Weimer et al., 2011).

Attempts to relate Joule heating to the polar cap index were reported by Chun et al. (1999, 2002). Basically they have derived hemispheric Joule heat production rates from the "Assimilative Mapping of Ionospheric Electrodynamics" (AMIE) procedure through a number of well documented storms. The AMIE procedure uses data from a large number of magnetometers, coherent and incoherent radars, digisondes and satellites with instruments for detection of electron precipitation and electric fields, to produce global patterns of various electrodynamic parameters. AMIE calculates Joule heating rate (jh) as shown in Eq. 17:

$$jh = \Sigma_P E^2 \tag{17}$$

where Σ_P is the ionospheric height-integrated Pedersen conductivity and E the electric field. The Joule heating may be integrated over either hemisphere or globally to produce the total Joule heating power. From the study by Chun et al. (1999), the total Joule heating power for the northern hemisphere (JH) was estimated for each season separately and compared to the corresponding values of the polar cap index, PCN. Their results are shown in Eqs. 18a-d:

Winter: $JH = 4.84\,PCN^2 + 16.9\,PCN + 5.6$ [GW]   (18a)

Equinox: $JH = 4.14\,PCN^2 + 25\,PCN + 8.9$ [GW]   (18b)

Summer: $JH = 14.39\,PCN^2 + 23.7\,PCN + 11.5$ [GW]   (18c)

All data: $JH = 4.03\,PCN^2 + 27.3\,PCN + 7.7$ [GW]   (18d)

Chun et al. (1999) noted that the expressions are valid also for negative values of the PCN index, which may occur during reverse convection cases, particularly during the summer season. Further, they compared the results of the above modelling (Eqs. 18a-d) with the simpler models based on either the cross polar cap voltage, Φ_{PC}, or the Auroral electrojet indices, AE, and found them about equally representative for the AMIE-derived Joule heating rates. This result is not so surprising since the cross polar cap potentials as well as the auroral electrojet indices are closely correlated with the polar cap indices as demonstrated in the previous sections.

From Thule PCN data and NOAA SPW Hemispheric Auroral Power (AP) data based on measurements from NOAA POES satellites through the years 1999-2002, and selecting only the northern passes, we have prepared the scatter plot displayed in Fig. 18. The heavy dots mark averages of AP values (in GW) through every unit of the PCN index. From Fig. 21 one may note the minimum in average AP around 10 GW at PCN = 0. The AP values are increasing both for increasing negative and positive PCN values but there is a marked difference between the slopes for positive and negative index values. The slopes for negative index values, particularly for the summer data, are much lower than those for the positive index data. For positive index values the average variations could be described by Eqs. 19a-c:

$$\text{Winter:} \qquad AP = 13.5\, PCN + 10 \qquad [GW] \qquad\qquad (19a)$$

$$\text{Equinox:} \qquad AP = 13.5\, PCN + 10 \qquad [GW] \qquad\qquad (19b)$$

$$\text{Summer:} \qquad AP = 11.5\, PCN + 10 \qquad [GW] \qquad\qquad (19c)$$

These results are not so different from those reported by Chun et al. (2002) taking into account that they did not discriminate against negative PCN index values, which they – looking at their data (their Fig. 9) - might well have done.

Summarizing from the above analyses, there appears to be quite consistent relations between the PC indices and the heating of the thermosphere by Joule heating as well as by

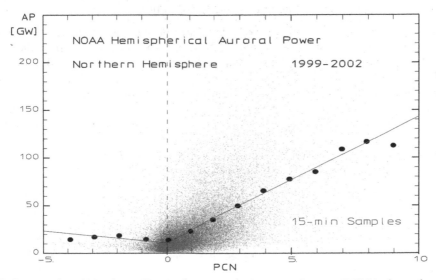

Fig. 18. Scatter plot of Northern Hemispheric Auroral Power data vs. PCN index values.

auroral particle precipitation. There are marked differences between cases with positive (forward convection) and negative (reverse convection) PC index values. Such differences are probably caused by the much reduced size of the active auroral and polar cap regions for the northward IMF Bz cases that causes the major part of the negative PCN values.

6.4 Relations between PC indices and the ring current

The Earth is encircled by currents flowing near equator at distances of typically 4-6 Re. These currents could be divided into the symmetrical part (RCS) that is formed all the way around the Earth, mostly by drifting mirroring energetic electrons and ions, and the partial ring current (RCP). Referring to Fig. 1, field-aligned Region 1 (R1) currents flow from the dawn magnetospheric boundary regions to the ionosphere at the morning side. Part of this current flows equatorward across the auroral region to feed the upward Region 2 (R2) field aligned currents (FACs) emerging from lower latitudes. The R2 FACs flow to the ring current region and add to the ring current at the night side to form a partial ring current. From the evening side of the ring current region, the R2 FACs flow to the ionosphere; part of these currents flows poleward crossing the auroral region and add to the R1 FACs leaving the ionosphere to end at the dusk boundary regions. It is also possible that a substantial part of the partial ring current is built from electrons and ions that simply drift directly between the magnetospheric dawn and dusk boundary regions across the near-tail region.

The ring current intensities are detected from a network of low-latitude magnetometer stations whose data are sent to World Data Centre WDC-C2 in Kyoto and processed to provide indices for the symmetrical as well as the partial ring currents (Sugiura & Kamei, 1981). The hourly symmetrical deflections scaled from the horizontal (H) components give the Dst index. The corresponding symmetrical index scaled from 1-min values of the H components provides the SYM-H index while the corresponding scaling of 1-min values of the D components generates the SYM-D index. Similarly the asymmetrical parts of the 1-min H and D components generate the ASY-H and ASY-D indices.

The asymmetric ring current index, ASY-H, have been provided by WDC-C2 (Iyemori et al., 2000) as 1-min values. For the present statistical study a less detailed time resolution, which reduces the inherent scatter in the data, is considered appropriate. Hence, the ASY-H and the polar cap indices, PCC, have been averaged to form 5-min samples. The two parameter sets have been subjected to linear correlation analysis using a stepwise variable delay between samples of the respective time series assuming that the most appropriate delay gives maximum value of the correlation coefficient. With this delay imposed on all pairs of samples of the time series, a linear relation between two parameter sets is found by least squares regression analysis. The average deviation, S0, the numerical deviation, S1, and the RMS standard deviation, S2, are calculated from the assumed linear relation.

Figure 19 displays ASY-H against PCC. The 15 min delay noted in the figure was found to provide optimum correlation (Rx=0.743). The regression line plotted in the diagram characterizes ASY-H against PCC. For this relation, the scatter parameters, S0, S1, and S2, are listed in the figure. The standard deviation (S2) is ~18 nT. A noteworthy feature in the display is the persistent linear relations of the average ASY-H values on PCC (Eq. 20) up to high levels

$$ASY\text{-}H = 12.1\,PCC + 11.5\,[nT] \tag{20}$$

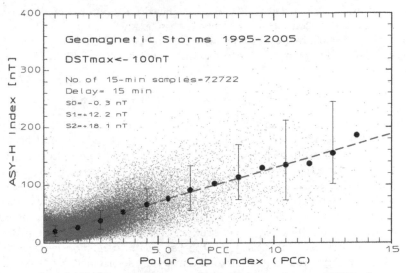

Fig. 19. Scatter plot of ASY-H partial ring current index vs. polar cap PCC index.

The Dst ring current index is closely related to the presence of energetic electrons and ions, notably H+ and O+, in the inner magnetosphere. Their different drift motion, positive ions westward and electrons eastward, in the geomagnetic dipole-like field contributes a net westward electric current. A succesful relation between the accumulated kinetic energy of the charged particles and the Dst* index is provided by the Dessler-Parker-Sckopke relation (Dessler & Parker, 1959; Sckopke, 1966) in Eq. 21:

$$Dst *[nT] = 4.0 \cdot 10^{-30} E_{RC} [keV] \qquad (21)$$

where Dst* is the Dst index corrected for the contributions from magnetopause currents (MPC) while E_{RC} is the total kinetic energy of particles trapped in the ring current region.

Thus, the Dst index represents the energy stored in the ring current. Hence the merging electric field, E_M, or, equivalently, the polar cap PC index should be considered to represent a source function for the Dst index rather than being related to its current value. Following Burton et al. (1975) the change in the Dst index with time could be written:

$$dDst*/dt = Q - Dst* /\tau [nT/hr] \qquad (22)$$

where Q in Eq. 22 is the source term while the last term is the ring current loss function controlled by the decay time constant τ here measured in hours. For the small actual MPC corrections, the Dst dependent statistical values provided in Joergensen et al. (2004) are used while the decay function given in Feldstein et al., (1984) is used for the loss term. This function uses two decay time constants, τ = 5.8 hrs for large disturbances where Dst < -55 nT, and τ = 8.2 hrs for small disturbances where Dst > -55 nT. Now, the relation in Eq. 22 has only terms relating to the source function Q and to the observed initial Dst index values.

In Burton et al. (1975) the source term Q is related to the Y_{GSM} component of the solar wind electric field. In the analysis by Stauning (2007), in addition to the dependence on E_M, the relation of Q to the polar cap index PCC was examined for a number of storm event cases during the interval 1995-2002. Here we repeat these analyses using selected large storm events through 1995-2005. In all these cases we first derive the temporal change at time t=T in the hourly Dst* index from the hourly values at t=T-1 and t=T+1 [hrs] by the simple differential term:

$$dDst^*/dt (T) = (Dst^*(T+1) - Dst^*(T-1))/2. [nT/hr] \qquad (23)$$

In order to derive the source function, Q, to be used in Eq. 22, the average slope values defined by Eq. 23 are corrected by adding the decay term defined above using the current Dst* value at t=T. The resulting source function, Q_{obs}, is then related to the potential source parameters with a variable positive or negative delay imposed on the relation. The parameters, the PC indices, are provided at a more detailed time resolution (typically 1-min) than the hourly source function values. By shifting the averaging interval by delays varying on minute scale, hourly averages of the parameters are correlated with the hourly source function values to derive the delay that produces the maximum correlation through the ensemble of storm events. With this delay the best fit linear relation between the source function values and the relevant source parameter is determined by linear regression analysis.

The scatter plots in Fig. 20 present the Dst* source function, Q_{obs} , based on observed hourly Dst values plotted against the combined polar cap index, PCC, as a potential source parameter. The relation between the best fit "equivalent" source function, Q_{eq}, and the source parameter values, PCC, is then expressed in a linear function:

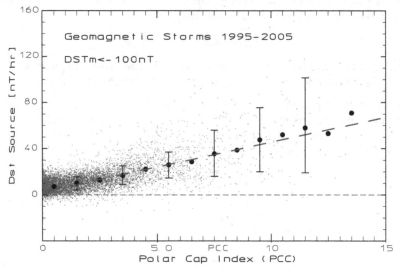

Fig. 20. Scatter plot of Dst* source function vs. polar cap PCC index.

$$Q_{eq} = 4.6 \, PCC + 1.2 \, [nT/h] \qquad (24)$$

with standard deviation equal to 9.4 nT/h.

With continuous time series of source parameter values, specification of the relational constants appearing in Eq. 24, and initial Dst values, it is now possible to integrate Eq. 22 to derive values of an "equivalent" Dst index, Dst$_{eq}$, through any interval of time. The procedure has been applied to the major geomagnetic storms (Dst<-100 nT) occurring through cycle 23. For each storm interval the calculations have been carried through intervals of 4 days. For each interval the equivalent Dst series was given an initial value equal to the observed Dst at the start of the first day, while all later Dst$_{eq}$ values through the 4 days were derived solely from the integration of Eq. 22 without attachment to observed Dst values.

Figure 21 displays an example for the case of a moderate geomagnetic storm on 7 – 10 April 1995. The peak (negative) Dst value is -150 nT observed at 1800 UT on 7.4.1995. In the figure the observed Dst values are displayed in black line with "dots" and located in the lower part of the diagram. The merging electric field, E$_M$, is displayed in thin green line in the upper half of the diagram. By its definition E$_M$ is non-negative. Values of the PCC index (also non-negative) are displayed in thin blue line with small squares while values of the asymmetric ring current index, ASY-H, are displayed in red line with crosses mainly in the upper part. Values of the equivalent Dst index calculated from the PCC indices are shown in blue line with squares.

The merging electric field, Em, and the asymmetric index, ASY-H have also been used as source functions similarly to the above derivation of the Dst indices using the PCC indices. Values of the equivalent Dst index calculated from E$_m$ are displayed in the thin green line in

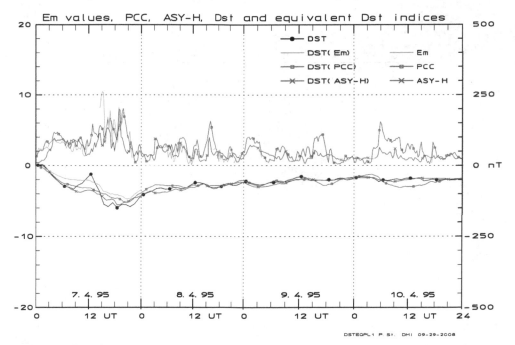

DSTEQPL1 P. St. DMI 09-29-2008

Fig. 21. Display of parameters E$_M$, PCC, ASY-H (in upper half of fields), and observed ring current index, Dst, and derived ring current indices Dst$_{eq}$(E$_M$), Dst$_{eq}$ (PCC), and Dst$_{eq}$ (ASY-H)

the lower part. In this case the series of equivalent Dst values based on E_m terminates after the first day since there is a break in the E_M data at midnight between 7 and 8 April 1995. The equivalent Dst indices using ASY-H in the source function are displayed in red line with crosses through all 4 days. In this case the PCC and the ASY-H index data are continuously available for the integration of Eq. 22 throughout the storm interval.

It is clear from Fig. 21 that there is good agreement between the observed and the equivalent Dst values throughout the 4 days of integration. A similar case is displayed in Fig. 22. The PCS index was not available for the storm. Hence, here the calculations were based solely on the PCN index series. Based on the peak values of the Dst index, this case illustrates one of the strongest geomagnetic storm events of cycle 23. Generally, going from the moderate to the strong storm cases gives less agreement between observed and equivalent Dst values. There is no clear indication in these plots whether the equivalent Dst values based on E_M are better or worse than those based on PCC, PCN, AL or ASY-H.

In order to resolve the question whether using E_M, PCC, PCN, AL or ASY-H in the source function give better agreement the average differences and variances between the observed and the equivalent Dst values were calculated for all cases studied. The results are listed in Table 3. It should be noted that the data sets forming the basis for Dst_{eq} calculations are not completely identical since the E_M data are often interrupted temporarily or non-existing for the storm cases. The reason for the lack of E_M data is often the disabling of the solar wind

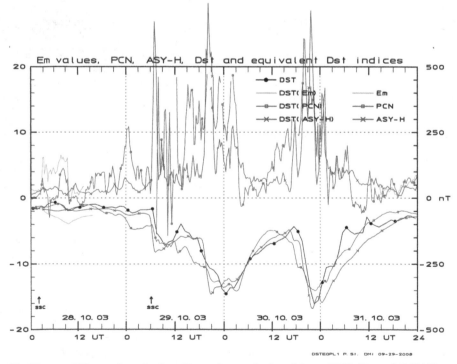

Fig. 22. Observed Dst and equivalent Dst values calculated from PCN, E_M and ASY-H, respectively, through a major magnetic storm of cycle 23.

detectors by strong high-energy solar proton fluxes. Hence the equivalent Dst values based on E_M data are often available during the first part of the storm interval only and the total number of samples is considerably reduced (cf. Table 3). Since the integration extending through 4 days becomes less and less certain with increasing interval length this preference for the first days may give the E_M-based Dst_{eq} values an advantage in the correlation with Dst over the PCC, PCN, AL or ASY-H-based values.

Parameter\Source	E_M field	PCC index	PCN index	AL index	ASY-H index
No. samples	3992	7776	8640	7968	8640
Avr. deviation	-7.6	1.8	2.7	11.5	- 4.9 nT
Std. deviation	27.6	30.6	31.4	34.2	24.7 nT

Table 3. Equivalent Dst versus observed Dst index through major geomagnetic storms 1995-2005

The best source function when comparing the average deviation between the observed Dst and the calculated equivalent values is provided by the PCC index series. Best to reduce the variance (standard deviation) is the ASY-H index followed by the solar wind merging electric field, E_M. The PCC index series gives larger variance than E_M but is continuously available in twice the number of cases. PCN based on Thule data are virtually available continuously. Thus, the PC indices, particularly the PCC index series, may serve to calculate the ring current indices, Dst, SYM-H, and ASY-H, in real time if the magnetic data from Thule and eventually Vostok are available on-line.

7. Conclusions

The polar caps provide terrestrial "windows" to the solar wind and enable monitoring of the merging electric field, E_M, the most important parameter for transfer of energy and momentum from the solar wind into the Earth's environment. The intruding solar wind energy is used to power a range of geophysical processes such as the large-scale electric potential patterns and the associated convection and current systems in the high-latitude magnetospheric and ionospheric regimes. The intensities of the transpolar ionospheric currents are very well suited to monitor the solar wind electric field. The associated magnetic disturbances at ground, when properly scaled to compensate for the regular daily and seasonal variations in conductivities, provide the polar cap index, PC, which can serve as a proxy for the merging electric field.

In addition, the PC indices, in particular when available from both polar caps, may serve to indicate the level of further geophysical disturbances such as polar ionospheric electric fields and plasma motions, auroral activities characterized, for instance, by the auroral electrojet indices (AL, AU, and AE), heating of the upper atmosphere by Joule dissipation (JH) and particle precipitation (AP), mid-latitude magnetic disturbances characterized by the Kp index, and the development of partial and symmetrical ring currents in the equatorial regions characterized by the Dst, the SYM-H, and the ASY-H indices. Such disturbances constitute an essential part of the conditions termed "Space Weather".

The polar cap indices are derived from a single station in each hemisphere, Thule in the northern polar cap and Vostok in the southern, both of which provide the basic magnetic data in high resolution and on-line in real time. These characteristics could be contrasted to those applicable to other indices used to characterize geophysical disturbances such as the auroral electrojet indices requiring data from 8-12 observatories to provide a reliable index, or the ring current indices based on 4-6 observatories. The estimation of Joule or particle heating power need either an extended network of magnetometer stations to run the AMIE procedure or polar satellite passes to derive the transpolar potential structure and the intensities of precipitating particles. Even in comparison with observations from interplanetary satellites, the monitoring of disturbances through the polar cap indices provides the advantage of not being affected by the intense high-energy particle radiation that often accompany the strong solar outburst, where the reliable monitoring of possible geophysical disturbances, that is, adverse space weather conditions, is particularly important.

8. Acknowledgments

The author is indebted to the observatory staffs for their dedicated efforts to provide magnetic data of utmost quality and continuity from the DMI geomagnetic observatory in Qaanaaq (Thule), Greenland, and the AARI geomagnetic observatory in Vostok at Antarctica. The author gratefully acknowledges the access to IMP8, Geotail, WIND and ACE satellite data on solar wind plasma and magnetic field observations provided through the NSSDC and the Caltech Research Laboratory data centres. We are also indepted to the NOAA SWP for supplying the NOAA POES satellite data. Furthermore, we gratefully acknowledge the access to Dst, SYM, ASY, and Auroral Electrojet indices from WDC-C2 at Kyoto University and express our deep appreciation of the efforts invested in the collection and processing of the geomagnetic data.

9. References

Akasofu, S.-I. (1979). Interplanetary energy flux associated with magnetospheric substorms, doi:10.1016/0032-633(79)90119-3, *Planet. Space Sci.*, 27, 425.

Araki, T. (1994). A physical model of geomagnetic sudden commencement, in: *Solar Wind sources of Magnetospheric Ultra-Low-Frequency Waves*, Geophys. Monogr. 81, eds.: M.J. Engebretson et al., pp. 183-200, AGU, Washington, D.C.

Borovsky, J. E., Layraud, B. & Kuznetsova, M. M. (2009). Polar cap potential saturation, dayside reconnection, and changes to the magnetosphere, *J. Geophys. Res.*, 114, A03224, doi:10.1029/2009JA014058.

Boyle, C. B., Reiff, P. H. & Hairston, M. R. (1997). Empirical polar cap potentials, *J. Geophys. Res.*, 102, 111.

Burton, R. K., McPherron, R. L., & Russell C. T. (1975). An empirical relationship between interplanetary conditions and Dst, *J. Geophys. Res.*, 80, 4204-4214.

Chun, F. K., Knipp,D. J., McHarg, M. G., Lu, G., Emery, B. A., Vennerstrøm, S. & Troshichev, O. A. (1999). Polar cap index as a proxy for hemispheric Joule heating, *Geophys. Res. Lett.*, 26 , 1101-1104.

Chun, F.K., Knipp, D. J., McHarg, M. G., Lacey, J. R., Lu, G., & Emery, B. A. (2002). Joule heating patterns as a function of polar cap index, *J. Geophys. Res.*, 107 (A7), doi:10.1029/2001JA000246.

Davis, T. N. & Sugiura, M. (1966). Auroral electrojet activity index AE and its Universal Time variations, *J. Geophys. Res.* 71, 785.

Dessler, A. J. & Parker, E. N. (1959). Hydromagnetic theory of geomagnetic storms, *J. Geophys. Res.*, 64, 2239-2259.

Fairfield, D. H. (1968). Polar magnetic disturbances and the interplanetary magnetic field, *COSPAR Space Research VIII*, 107.

Farris, M.H. & Russell, C. T. (1994). Determining the standoff distance of the bow shock: Mach number dependence and use of models, *J. Geophys. Res.*, 99, 17681-17689.

Feldstein, Y. I., Pisarsky, V. Yu, Rudneva, N. M. & Grafe, A. (1984). Ring current simulation in connection with interplanetary space conditions, *Planet. Space Sci*, 32, 975-984.

Fiori, R. A. D., Koustov, A. V., Boteler, D. & Makarevich, R. A. (2009). PCN magnetic index and average convection velocity in the polar cap inferred from SuperDARN radar measurements, *J. Geophys. Res.* 114, A07225, doi:10.1029/2008JA013964.

Friis-Christensen, E. & Wilhjelm, J. (1975). Polar cap currents for different directions of the interplanetary magnetic field in the Y-Z plane, *J. Geophys. Res.*, 80, 1248-1260.

Fukushima, N. (1976). Generalized theorem for no ground magnetic effect of vertical currents connected with Pedersen currents in the uniform-conductivity ionosphere, *Rep. Ionos. Space Res. Jpn.*, 30, 35.

Greenwald, R. A., Bristow, W. A., Sofko, G. J., Senior, C., Ceriser, J.-C. & Szabo, A. (1995). SuperDual Auroral Radar Network radar imaging of dayside high-latitude convection under northward interplanetary magnetic field: Toward resolving the distorted two-cell versus multicell controversy, *J. Geophys. Res.*, 100, 19.661.

Hairston, M. R., Heelis, R. A., & Rich, F. J. (1998), Analysis of the ionospheric cross polar cap potential drop using DMSP data during the National Space Weather Program study period, *J. Geophys. Res.*, 103, 26,337.

Hairston, M. R., Drake, K. A., and Skoug, R. (2005). Saturation of the ionospheric polar cap potential during the October-November 2003 superstorms, *J. Geophys. Res.*, 110, A09826, doi:10.1029/2004JA010864.

Huang, C.-S. (2005). Variations of polar cap index in response to solar wind changes and magnetospheric substorms, *J. Geophys. Res.*, 110, A01203, doi:10.1029/2004JA10616.

Iijima, T. & Potemra, T. A. (1976a). The amplitude distribution of field-aligned currents at northern high latitudes observed by Triad, *J. Geophys. Res.*, 81, 2165-2174.

Iijima, T. & Potemra, T. A. (1976b). Field-aligned currents in the dayside cusp observed by Triad, *J. Geophys. Res.*, 81, 5971.

Iyemori, T., Araki, T., Kamei, T. & Takeda, M. (2000). Mid-latitude geomagnetic indices "ASY" and "SYM" for 1999, *Geomagnetic indices home page*, ed.: T. Iyemori, (http://swdcwww.kugi.kyoto-u.ac.jp/dstdir/index.html), WDC-C2 for Geomagnetism, Kyoto University.

Janzhura, A, Troshichev, O. A. & Stauning, P. (2007). Unified PC indices: Relation to isolated magnetic substorms, *J. Geophys. Res.*, 112, A09207, doi:10.1029/2006JA012132.

Jorgensen, A. M., Spence, H. E., Hughes, W. J. & Singer, H. J. (2004). A statistical study of the ring current, *J. Geophys. Res.*, 109, A12204, doi:10.1029/2003JA010090.

Kan, J. R. & Lee, L. C. (1979). Energy coupling function and solar wind-magnetosphere dynamo, *Geophys. Res. Lett.*, 6, 577.

Kuznetsov, B. M. and Troshichev, O. A. (1977). On the nature of polar cap magnetic activity during undisturbed periods, *Planet. Space Sci.*, 25, 15–21, 1977.

Lukianova, R., Troshichev, O. A. & Lu, G. (2002). The polar cap magnetic activity indices in the southern (PCS) and northern (PCN) polar caps: Consistency and disrepancy, Geophys. Res. Lett., 29(18), 1879, doi:10.1029/2002GL015179.

Lukianova, R. (2003). Magnetospheric response to sudden changes in the solar wind dynamic pressure inferred from polar cap index, J. Geophys. Res., 108 (A12), 1428, SMP-11. doi:10.1029/2002JA009790.

Mansurov, S. M. (1969). A new evidence of connection between Space and Earth's magnetic fields, Geomag. and Aeron., 9, 768.

Newell, P. T. & Meng, C.-I. (1994). Ionospheric projections of magnetospheric regions under low and high solar wind pressure conditions, J. Geophys. Res., 99 (A1), 273-286.

Reiff, P. H., Spiro, R. W. & Hill, T. W. (1981). Dependence of polar cap potential drop on interplanetary parameters, J. Geophys. Res., 86, 7639.

Rich, F. J., & Hairston, M. (1994). Large-scale convection patters observed by DMSP, J. Geophys. Res., 99, 3827.

Sckopke, N. (1966). A general relation between the energy of trapped particles and the disturbance field near the Earth, J. Geophys. Res., 71, 3125-3130.

Shue, J.-H., Chao, J. K., Fu, H. C., Russell, C. T., Song, P., Khurana, K. K. & Singer, H. J. (1997). A new functional form to study the solar wind control of the magnetopause size and shape, J. Geophys. Res., 102, 9497-9511.

Siscoe, G. L., Crooker, N. U. & Siebert, K. D. (2002). Transpolar potential saturation: Roles of region 1 current system and solar wind ram pressure, J. Geophys. Res. 107 (A10), 1321, doi:10.1029/2001JA009176.

Sonnerup, B. (1974). Magnetopause reconnection rate, J. Geophys. Res., 79, 1546-1549.

Spreiter, J. R., Summers, A. L. & Alksne, A. Y. (1966). Hydromagnetic flow around the magnetosphere, Planet. Space Sci., 14, 223-253.

Stauning, P., Troshichev, O. A. & Janzhura, A. (2006). Polar Cap (PC) index. Unified PC-N (North) index procedures and quality. DMI Scientific Report, SR-06-04. (available at www.dmi.dk/publications/sr06-04.pdf).

Stauning, P., (2007). A new index for the interplanetary merging electric field and geomagnetic activity: Application of the unified polar cap indices, Space Weather, 5, S09001, doi:10.1029/2007SW000311.

Stauning, P. & Troshichev, O. A. (2008). Polar cap convection and PC index during sudden changes in solar wind dynamic pressure, J. Geophys. Res., 113, doi:10.1029/2007JA012783.

Stauning, P., Troshichev, O. A. & Janzhura, A. (2008). The Polar Cap (PC) index:. Relations to solar wind parameters and global activity level, J. Atmos. Solar-Terr. Phys., doi:10.1016/j.jastp.2008.09.028.

Stauning, P. (2011), Determination of the quiet daily geomagnetic variations for polar regions, J. Atmos. Solar-Terr. Phys., doi:10.1016/j.jastp.2011.07.004.

Sugiura, M. & Davis, T. N. (1966). Auroral electrojet activity index AE and its universal time variations, J. Geophys. Res., 71, 785-801.

Sugiura, M. & Kamei, T. (1981). Description of the hourly Dst index, in: Geomagnetic indices home page, ed.: T. Iyemori, (http://swdcwww.kugi.kyoto-u.ac.jp/dstdir/index.html),   WDC-C2   for Geomagnetism, Kyoto University.

Svalgaard, L. (1968). Sector structure of the interplanetary magnetic field and daily variation of the geomagnetic field at high-latitudes, *Scientific Report, R-8*, Danish Meteorological Institute.

Svalgaard, L., (1972). Interplanetary magnetic sector structure 1926-1971, *J. Geophys. Res.*, 77, 4027.

Svalgaard, L., (1973). Polar cap magnetic variations and their relationship with the interplanetary magnetic sector structure, *J. Geophys. Res.*, 78, 2064-2078.

Tobiska, W. K., Bouwer, S. D. & Bowman, B. R. (2008). The development of new solar indices for use in thermospheric density modelling, *J. Atmos. Solar-Terr. Phys.*, 70, 803-819.

Troshichev, O. A., Dmitrieva, N. P. & Kuznetsov, B. M. (1979). Polar Cap magnetic activity as a signature of substorm development, *Planet. Space Sci.*, 27, 217.

Troshichev, O. A. & Andrezen, V. G. (1985). The relationship between interplanetary quantities and magnetic activity in the southern polar cap, *Planet. Space Sci.*, 33, 415.

Troshichev, O. A., Kotikov, A. L., Bolotinskaya, B. D., Andrezen, V. G. (1986). Influence of IMF azimuthal component on magnetospheric substorm dynamics, *J. Geomagn. Geoelectr.*, 38, 1075.

Troshichev, O. A., Andrezen, V. G., Vennerstrøm, S. & Friis-Christensen, E. (1988). Magnetic activity in the polar cap – A new index, *Planet. Space Sci.*, 36, 1095.

Troshichev, O., Hayakawa, H., Matsuoka, A., Mukai, T. & Tsuruda, K. (1996). Cross polar cap diameter and voltage as a function of PC index and interplanetary quantities, *J. Geophys. Res.*, 101, 13,429.

Troshichev, A. O., Lukianova, R. Y., Papitashvili, V. O., Rich, F. J. & Rasmussen, O. (2000). Polar Cap index (PC) as a proxy for ionospheric electric field in the near-pole region, *Geophys. Res. Lett.*, 27, 3809.

Troshichev, O. A., Janzhura, A. & Stauning, P. (2006). Unified PCN and PCS indices: method of calculation, physical sense and dependence on the IMF azimuthal and northward components, *J. Geophys. Res.*, 111, A05208, doi:10.1029/2005JA011402.

Troshichev, O. A., Janzhura, A. & Stauning, P. (2007). Magnetic activity in the polar caps: Relation to sudden changes in the solar wind dynamic pressure, *J. Geophys. Res.*, 112, A11202, doi10.1029/2007JA012369.

Vennerstrøm, S. (1991). The geomagnetic activity index PC, PhD Thesis, *Scientific Report 91-3*, Danish Meteorological Institute, 105 pp.

Vennerstrøm, S., Friis-Christensen, E., Troshichev, O. A. & Andrezen, V. G. (1991). Comparison between the polar cap index PC and the auroral electrojet indices AE, AL and AU, *J. Geophys. Res.*, 96, 101.

Weimer, D. R. (2005). Improved ionospheric electrodynamics models and application to calculating Joule heating rates, *J. Geophys. Res.*, 110, A05306, doi:10.1029/2004JA010884.

Weimer, D. R. & King, J. H. (2008). Improved calculations of interplanetary magnetic field phase front angles and propagation time delays, *J. Geophys. Res.*, 113, A01105, doi:10.1029/2007JA012452.

Weimer, D. R., Clauer, C. R., Engebretson, M. J., Hansen, T. L., Gleisner, H., Mann, I. & Yumoto, K. (2010). Statistical maps of geomagnetic perturbations as a function of the interplanetary magnetic field, *J. Geophys. Res.*, 115, A10320, doi:10.1029/2010JA015540.

Weimer, D. R., Bowman, B. R., Sutton, E. K. & Tobiska, W. K. (2011). Predicting global
average thermospheric temperature changes resulting from auroral heating, *J. Geophys. Res.*, 116, A01312, doi:10.1029/2010JA015685.

Wilhjelm, J., Friis-Christensen, E., Potemra, T. A. (1972). The relationship between
ionospheric and field-aligned currents in the dayside cusp, *J. Geophys. Res.* 83, 5586.

5

Sudden Impulses in the Magnetosphere and at Ground

U. Villante and M. Piersanti

Dipartimento di Fisica, Università e Area
di Ricerca in Astrogeofisica, L'Aquila
Italy

1. Introduction

Sudden Impulses (SI) are rapid variations of the magnetospheric and geomagnetic field which are usually related to the Earth's arrival of sudden increases in the dynamic pressure of the solar wind (SW), generally associated with interplanetary shock waves or discontinuities. Such impulses often precede geomagnetic storms: in this case they are referred as Storm Sudden Commencements (SSC). Incoming SW pressure pulses compress the magnetosphere, increase the magnetopause and tail currents, and possibly other magnetospheric/ionospheric current systems as well; correspondingly, the magnetospheric and ground fields generally increase to a new state over about a two to fifteen minute period. The field variation associated to SI mostly occurs along the north/south component (B_z in the magnetosphere, H in the geomagnetic field). Additional contributions also come from the effects of the tail current (mostly in the nightside sector) and from those of the ring current (during more active magnetospheric conditions). The further development of field aligned currents (FAC) and ionospheric currents typically makes the SI manifestation at ground much more complex than in the magnetosphere: on the other hand, since earliest investigations (Matsushita, 1962; Nishida and Jacobs, 1962), it is well known that different transient waveforms are detected at different ground stations. Given their global simultaneous occurrence and clear onset time, SI provide a good opportunity for understanding the transient response of the magnetosphere and ionosphere to the SW variations.

In the present paper we review some aspects of the SI manifestations, as they are observed in the magnetosphere (basically at geosynchronous orbit, ≈ 6.6 Re, Re being the Earth radius) and at ground. For previous reviews of the experimental and theoretical aspects of SI events the reader is referred to Matsushita (1962), Nishida and Jacobs (1962), Siscoe et al. (1968), Nishida (1978), Smith et al. (1986), Araki (1994), Tsunomura (1998).

2. The SI manifestation at geosynchronous orbit

2.1 The general aspects and the local time dependence

Figure 1 (after Villante and Piersanti, 2011) shows the aspects of the SI manifestation at different LT (LT being the local time) along the geosynchronous orbit and the relationship between the magnetospheric field change (ΔB, B being the total field) and the increase of the

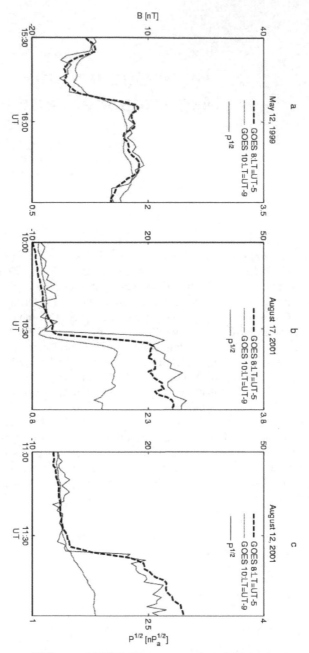

Fig. 1. (after Villante and Piersanti, 2011). Three examples of SI events at geosynchronous orbit. In each panel we show: the square root of the SW dynamic pressure (WIND spacecraft) and the magnetic field amplitude at GOES8 and GOES10. WIND data have been shifted to account for the SW propagation time

square root of the SW pressure $\Delta P^{1/2}$. As shown in panel a, the field jump is typically sharp in the entire dayside sector and basically reflects the behaviour of the change of the SW pressure, especially in the subsolar region. The ratio $R=\Delta B/\Delta P^{1/2}$ (hereafter referred as "relative response") at dawn is somewhat smaller (and smoother) than in the noon region ($R\approx21$ nT/nPa$^{1/2}$ and $R\approx25$ nT/nPa$^{1/2}$, respectively for the case in panel a). Such response attains smaller values in the dark hemisphere ($R\approx12$ nT/nPa$^{1/2}$ at $\approx1:30$ LT, $R\approx17$ nT/nPa$^{1/2}$ at $\approx5:30$ LT, for the case in panel b). Lastly, panel c shows that, in some cases, a continuous, small amplitude increase of the magnetospheric field is detected in the night sector, even in presence of a sharp change at dawn.

Although comparisons among different investigations are made ambiguous by the different criteria adopted for the selection of events, for the definition of "magnetospheric response", and for the large variety in its amplitude (and characteristics) in any time sector, an explicit LT modulation of such response has been extensively reported in the scientific literature. On the other hand, since the magnetopause current is mostly enhanced in the dayside sector during the magnetospheric compression, while the enhancement of the tail current produces a negative variation of the magnetospheric field, a strong day/night asymmetry might be expected in the SI manifestation at geosynchronous orbit. Consistently, Patel and Coleman (1970) reported that SI events in the nightside had smaller amplitude than in the dayside. Kokubun (1983) and Kuwashima and Fukunishi (1985) found that the magnetospheric change had highest values at local noon and very small values, or even negative, near midnight. More recently, Lee and Lyons (2004) concluded that SW pressure enhancements generally lead to a magnetospheric compression at all time sectors (with few exceptions in the nightside): it is strongest near noon and decreases toward dawn and dusk. Borodkova et al. (2005, 2006) found that, in general, the changes of the SW pressure (positive and negative) were associated with corresponding variations of the magnetospheric field; they also remarked that all the events without an explicit response were located either before 7:30 LT or after 16:30 LT. Wang et al. (2007) confirmed that the field variation peaked near local noon and decreased toward dawn and dusk. Figure 2a (after Villante and Piersanti, 2011) compares the amplitude of the geostationary response for events simultaneously detected at different LT: it clearly confirms a large data spread of the relative response in any time sector, together with an explicit LT modulation, with greater values at satellite located closer to the noon meridian; as can be seen, negligible and even negative magnetospheric responses are often detected in the dark sector.

2.2 The role of the SW parameters

As for other aspects of the magnetospheric dynamics, particular attention has been dedicated to the possible role of the North/South component of the interplanetary magnetic field ($B_{z,IMF}$). As a matter of fact, Sanny et al. (2002) showed that the variability of the magnetospheric field strength near local noon was independent on the IMF orientation but strongly influenced by changes of the SW pressure. Consistently, Wang et al. (2007) concluded that the IMF orientation does not affect the geosynchronous response significantly (see also Figure 2a). Kuwashima and Fukunishi (1985) and, more recently, Lee and Lyons (2004) suggested that midnight events associated with southward IMF orientations were often characterized by a dipolarization-like change similar to that one occurring during substorms, while the dayside response was mostly compressional; for northward IMF, a compression of the entire magnetosphere was generally observed, with few

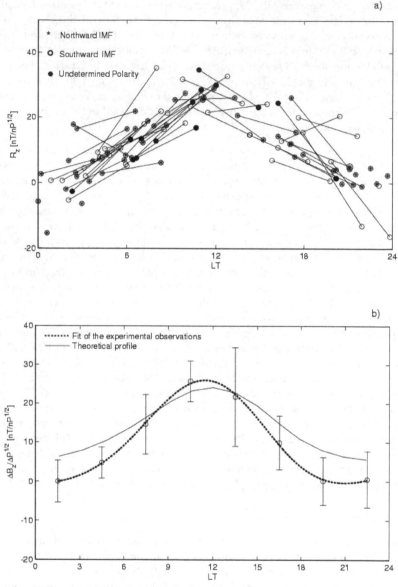

Fig. 2. a) (after Villante and Piersanti 2008) The relative response $R_z = \Delta B_z/\Delta P^{1/2}$ vs. local time (observation of the same event from two spacecraft are connected by a line. Empty circles identify events associated with Southward IMF; stars identify events associated with Northward IMF; black circles identify events associated with undetermined IMF polarity; b) (after Villante and Piersanti 2008) A comparison between average values of $\Delta B_z/\Delta P^{1/2}$ in each 3-h interval and the theoretical profile (solid line), determined considering the magnetic effects of the magnetopause current at geosynchronous orbit. The dotted line represents the fit of experimental measurements.

cases of depression near midnight. Lee and Lyons (2004) suggested that the magnetosphere is very sensitive to small SW pressure enhancements when the IMF is strongly southward for a long period of time. Focusing attention on the midnight sector, Wang et al. (2009) concluded that ≈75% of the negative responses were associated with southward IMF orientations. By contrast, Sun et al. (2011) revealed that the occurrence of positive or negative responses in the midnight sector had no obvious association with the sign of $B_{z,IMF}$, provided that no inversion of $B_{z,IMF}$ exists across the front of the impinging discontinuity.

2.3 The comparison with theoretical models

Kokubun (1983) evaluated the role of the magnetopause and tail currents and concluded that the geosynchronous responses were ≈30% smaller than expected. Figure 2b (after Villante and Piersanti, 2008), compares the 3-hr averages of R_z with the theoretical profiles evaluated assuming that the magnetic field change is determined by the transition between two steady states of the magnetosphere under different SW pressure conditions (i.e. different magnetopause currents, Tsyganenko, 2002a, 2002b). As a matter of facts, the observed average values reveal in the central part of the day a close correspondence with the predicted responses, a feature confirmed by an analysis of individual events (Villante and Piersanti, 2008). It suggests that, in this region, the field jumps are basically determined by the changes of the magnetopause current alone. The occurrence of negligible and negative responses makes the average values smaller than predicted in the dark region. It is worth noting, however, that, even in this region, the positive ΔB_z often show a substantial correspondence with the values predicted for the magnetopause current, especially for higher SW pressure jumps (Villante and Piersanti, 2011): it suggests that the dominant effects of the magnetopause current (i.e. the magnetospheric compression) might extend to a significant portion of the dark magnetosphere. On the other hand, the occurrence of negative variations in the nightside region (unpredictable in terms of the magnetopause current alone, solid line in Figure 2b) reveals, in several cases, a significant role of additional current systems. The interpretation of such events, however, requests a case by case analysis, paying attention to the SW and magnetospheric conditions in the period of interest (Villante and Piersanti, 2011). As a matter of facts, several approaches have been adopted to interpret the characteristics of the SI manifestation in the dark magnetosphere. Interesting results have been recently provided by Sun et al. (2011) who performed a MHD simulation of the nightside response to interplanetary shocks and concluded that when a shock sweeps over the magnetosphere, there exist mainly two regions: a positive response region caused by the compressive effect of the shock and a negative response region which is probably associated with the temporary enhancement of earthward convection in the nightside magnetosphere. In addition, according to their conclusions, a southward IMF would lead to a stronger and larger negative response region, and a higher shock speed would result in stronger negative and positive response region.

3. The SI manifestation at ground

3.1 The general aspects and the local time dependence

As previously underlined, the manifestation of ground SI is, in general, more complex than at geosynchronous orbit. This is because secondary effects like FAC and ionospheric currents contribute significantly in addition to the primary effect of the magnetopause current. On the other hand, it is now clear that FAC may modify the SI field even at middle and low latitudes (Kikuchi et al., 2001; Araki et al., 2006). As a matter of facts, shape and amplitude of the H waveform are strongly dependent on latitude and LT. Basically (Araki, 1994), at auroral latitudes, the waveform consists of two successive pulses with opposite sense (PI and MI, with typical duration of ≈1-2 min and ≈5-10 min, respectively). In the morning a positive pulse precedes and a negative pulse follows. The sense of the pulses is reversed in the afternoon and their amplitude decreases with decreasing latitude (Figure 3). Typically, SI manifestations are more simple at low latitudes, i.e. far from major high latitude and equatorial current systems. Here, the H behaviour becomes more step-like, but a two pulse structure with reduced amplitude is still identified. Figure 4 (after Villante and Piersanti, 2011) compares the aspects of ground events at $\lambda\approx36°$ (λ being the magnetic latitude) with the SW and geosynchronous observations. It confirms that, independently on LT, the low latitude response of the H component often consists of a simple monotonic increase with amplitude (ΔH) often comparable (although smaller) with the field jump observed at geosynchronous orbit. Interestingly, in any time sector, the H variations are accompanied by explicit negative variations of the D component (ΔD, perpendicular to H in the horizontal plane). In some cases, moreover, even at low latitudes, the main field jump is preceded by transient phenomena which do not appear in the magnetosphere: for example, the event in panel c is characterized by the occurrence of a preliminary positive impulse (PPI).

As for geosynchronous orbit, specific attention has been addressed to the LT dependence of the ground response. In a pioneering investigation, Ferraro and Unthank (1951) showed that, at low and middle latitudes, ΔH was larger near midnight than in daytime. More recently, Tsunomura (1998), who analysed events from low to middle latitudes ($\lambda\approx21°$-$43°$), determined a smaller ΔH during local morning with respect to the rest of the day. Araki et al. (2006, 2009) examined the LT dependence of the average ΔH at $\lambda\approx35.4°$, separately for summer and winter. They determined, in both seasons, a maximum near midnight, a minimum at 7-8 LT, and a secondary maximum on the dayside; in addition, the amplitude of the LT modulation was much larger in the summer (approximately by a factor ≈3). Russell et al. (1992, 1994a, b) reported that at low and middle latitudes ΔH is maximum around noon during northward IMF conditions; in addition, ΔH was found to decrease in the daytime sector and to enhance significantly in the night time sector in the case of southward IMF. At $\lambda\approx36°$, Francia et al. (2001) revealed a LT dependence of the relative response, $R_H=\Delta H/\Delta P^{1/2}$, characterized by a depressed value in the morning (≈10 nT/(nPa)$^{1/2}$), a greater amplitude in the evening and night sector (≈14-17 nT/(nPa)$^{1/2}$), and a maximum after the local noon (≈20 nT/(nPa)$^{1/2}$). Similarly, at subauroral latitudes ($\lambda\approx54°$-$58°$), the geomagnetic response showed strongly depressed values in the morning and enhanced values in the afternoon (≈30 nT/(nPa)$^{1/2}$; Russell and Ginskey, 1995). Recently, Shinbori et al. (2009) conducted a statistical analysis of the relative amplitude from middle ($\lambda\approx45°$) to equatorial latitudes ($\lambda\approx15°$) suggesting, between $\lambda\approx45°$ and $\lambda\approx36°$, a strong dawn-dusk asymmetry, with minimum and maximum values in the morning (8-9 LT) and afternoon (15- 17 LT), respectively, and some evidence for a new enhancement in the night time sector (20–03 LT). They also showed that the amplitude of the relative response does not vary under specific IMF conditions (northward/southward), suggesting negligible effects from the ring and tail current. As a matter of fact, the emerging overview reveals a LT dependence of the ground response (with amplitude dependent on latitude and season) significantly different than in the magnetosphere, revealing explicit contributions from FAC and ionospheric currents: basically, from low to high latitudes, it is characterized by smaller (even negative, at higher latitudes) values in the morning and greater values in the post-noon and midnight sector, while, at equatorial latitudes, it shows a strong enhancement around 11 LT.

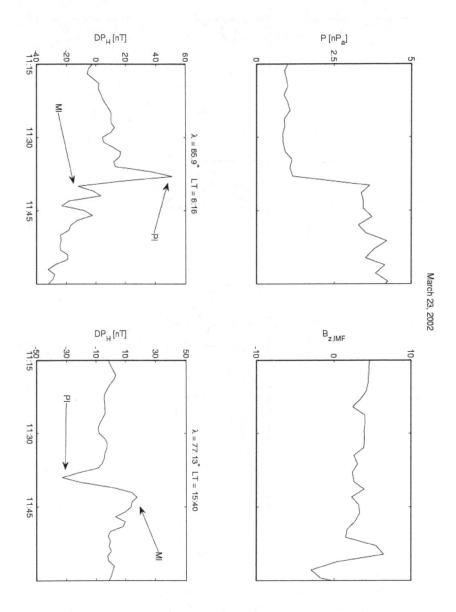

Fig. 3. Two examples of SI events at auroral latitudes in the local morning (right panel) and in local afternoon (left panel).

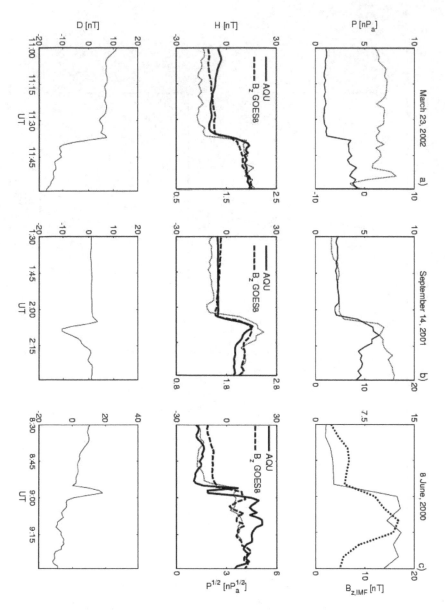

Fig. 4. (after Villante and Piersanti 2008) Three examples of SI events at low latitudes. Top panel: the SW dynamic pressure (solid line); the North/South component of the interplanetary magnetic field (dotted line). Central panel: the geomagnetic field component H at AQU; the magnetic field component B_Z at geosynchronous orbit; the square root of the SW pressure. Bottom panel: the geomagnetic field component D at AQU. WIND data have been shifted to account for the SW propagation time.

3.2 The comparison with theoretical models

The current understanding relates such complex scenario to the combined effects of the magnetospheric and ionospheric current systems. Namely, the total disturbance field (D_{SI}) of the H component is decomposed in different subfields (Araki, 1977, 1994; Araki et al., 1997, 2009):

$$D_{SI} = DL_{MI} + DP_{PI} + DP_{MI}$$

According to models, the direct effect of the increased magnetopause current propagates to low and middle latitudes as a compressive wave and produces a step like increase of the H component (DL_{MI} field, where L stands for low latitudes); its amplitude is largest at the equator and decreases with increasing latitude. A dusk-to-dawn electric field along the compressional wave front induces a twin ionospheric vortex system that produces a preliminary impulse of polar origin (DP_{PI}). The DP_{PI} field manifests as a preliminary reverse impulse (PRI) simultaneously observed at auroral latitudes in the afternoon and near the dip equator on the dayside, as well as a preliminary positive impulse (PPI) observed at auroral latitudes in the morning. On the other hand, if the increased pressure behind the SW discontinuity is kept up, the magnetospheric convection has to adjust itself to the compressed state of the magnetosphere: as a final result, it produces a twin polar vortex system (DP_{MI}) which is opposite to the DP_{PI} field and corresponds to the MI. Such DP_{MI} field is basically driven by an electric field originated in the polar region and transmitted from the outer magnetosphere through FAC which flow into the ionosphere in the morning side and away in the afternoon side. In this scheme, the preliminary impulse is exclusively due to current systems of polar origin, DP_{PI}, whereas the main impulse is due to the combined effect of DL_{MI} and DP_{MI}. On the other hand, Kikuchi et al. (2001) found that preliminary positive pulses tend to appear in the afternoon middle latitudes, and proposed that the generation mechanism is the magnetic effect of the FAC which are accompanied with the dusk-to-dawn electric fields.

The comparison between measurements at ground with those obtained by low altitude satellites above the ionosphere provided important insights on the ionospheric currents. For example, Araki et al. (1984), comparing ground and MAGSAT observations, showed that both H and D components showed variations with opposite sense at satellite and ground, revealing the existence of ionospheric currents associated with SI (at least near dawn and dusk, due to the orbital configuration). Han et al. (2007) examined Oersted and ground data and observed, in the night sector, very similar waveforms above and below the ionosphere; they then concluded that the ionospheric currents do not contribute significantly to nightside SI which, according to their conclusions, were dominantly caused by the enhanced magnetopause currents. By contrast, the waveforms observed by Oersted on the dayside were apparently different from those observed on the ground, reflecting the role of ionospheric currents. Corresponding to the PRI and the MI observed in the H component at the dayside dip equator, Oersted always observed an increase and a clear decrease in the magnetic field, respectively. These observational results suggest that the PRI at the dayside dip equator corresponds to a westward ionospheric current, and an eastward current is excited after the PRI. More recently, a comprehensive analysis of events simultaneously observed at ground and by CHAMP (Luhr et al., 2009) confirmed that night time events at ground are not (or minimally) affected by ionospheric currents. More in general, this analysis also showed that at latitudes smaller than $\lambda \approx 40°$ the amplitude of the field variation was practically the same at ground and satellite; a progressive latitudinal increase of the SI amplitude was determined at higher latitudes both on the ground and at satellite, suggesting the effects of FAC rather than those of currents flowing in the ionosphere.

Figure 5a (after Villante and Piersanti, 2011) shows the 3-hr average values of the ground relative responses $<R_H>$ at $\lambda \approx 36°$: as can be seen, the LT modulation is much less pronounced than at geostationary orbit. The average responses in Figure 5a are compared with the theoretical profiles expected for the magnetopause current alone (R_{CF}) and for the global magnetospheric current system (from the magnetopause, tail and ring current, B_T), as evaluated at the winter and summer solstice. Such comparison shows that the observed $<R_H>$ fall, in general, within the limits of the expected profiles from approximately premidnight up to noon (an aspect confirmed by an analysis of single events), suggesting a poor contribution from FAC and ionospheric currents on this component. In the post-noon region, the observed responses overcome the theoretical profiles on average by $\approx 50\%$ (and occasionally by a factor $\approx 2-3$ in individual cases, Villante and Piersanti, 2011), revealing explicit effects from the additional ionospheric currents in this time sector. Consistently, Shinbori et al. (2009) interpreted such LT (and latitudinal) dependence as the manifestation of superimposed ionospheric currents: in particular, at low and middle latitudes, the observed pattern would be related to currents (producing negative and positive variations of the H component in the dawn and dusk sector, respectively) generated by the enhanced dawn-to-dusk electric field (accompanying FAC) due to the compression of the magnetosphere. Note, in addition, that, the negative responses of the D component ($<R_D>$, figure 5b) are far from the theoretical profiles, suggesting an explicit influence (through the entire day) on this component of the additional current systems.

4. The ULF waves occurrence

SI manifestations are occasionally accompanied by trains of almost monochromatic ULF waves ($f \approx 1-10$ mHz, usually referred as "Pc5 pulsations", Villante, 2007) which manifest soon after the main variation and persist for few cycles. According to Zhang et al. (2009), at geosynchronous orbit, the magnitude of ULF waves associated to variations of the SW pressure is larger around noon than at dawn and dusk. Amata et al. (1986) showed that magnetospheric oscillations excited by a SW shock have a complicated structure: a magnetosonic wave propagating from the dayside to the nightside sector suddenly appears and resonant shear Alfvén waves are generated during the propagation of this magnetosonic wave. As regard to ground observations, Ziesolleck and Chamalaun (1993) found wave amplitudes increasing with latitude, while their frequency was not dependent on either latitude or longitude.

A spectacular example of SI-related waves is shown in Figure 6 (after Piersanti et al., 2012) in which, at GOES 8 position ($\approx 04:15$ LT), the SI manifestation itself practically consists in the onset of a large amplitude wave mode at $f=3.3$ mHz; by contrast, GOES 10 ($\approx 00:15$ LT) did not observe any similar wave activity in the midnight sector. Waves at the same frequency were detected at all ground stations (Figure 7 and Figure 8; after Piersanti et al., 2012), suggesting an interpretation in terms of a global oscillation mode of the whole magnetospheric cavity. Moreover, the characteristics of the wave amplitude and the behaviour of the polarization pattern suggest that, due to the variable length of the field line through the day, the magnetospheric wave leaked energy to field line resonance at $\lambda \sim 66°$ in the morning sector and at $\lambda \sim 71°$ in the noon sector.

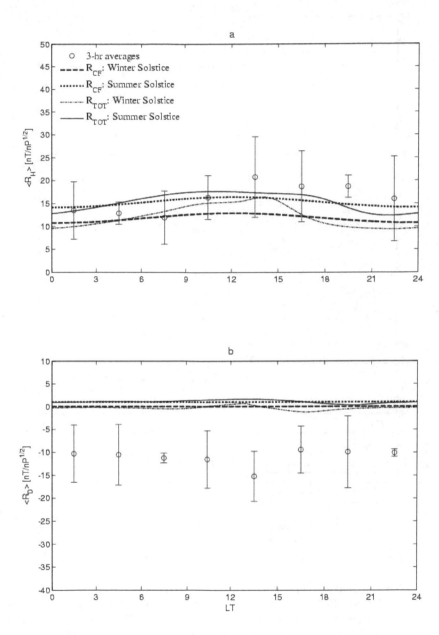

Fig. 5. (after Villante and Piersanti 2008) A comparison between averages values <R_H> in 3-hr intervals and the theoretical profiles determined considering the ground effects of the magnetopause current (B_{CF}-field) and those of the total current system (B_{TOT}-field). Both profiles have been evaluated at winter and summer solstice. b) The same for <R_D>.

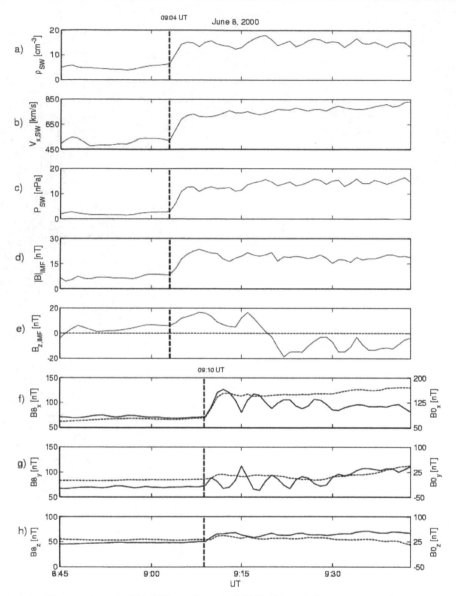

Fig. 6. (after Piersanti et al., 2011) The solar wind (WIND) and the magnetospheric observations (GOES) for the SI event occurred on June 8, 2000: a) the SW density; b) the SW velocity; c) the SW dynamic pressure; d) the IMF $|B|$; e) the IMF Bz component; f) The B_X component of the magnetospheric field measured by GOES8 (solid line) and GOES10 (dashed line); g) the B_Y component of the magnetospheric field measured by GOES8 (solid line) and GOES10 (dashed line); The B_Z component of the magnetospheric field measured by GOES8 (solid line) and GOES10 (dashed line). The vertical dashed lines identify the SW pressure jump and the SI occurrence.

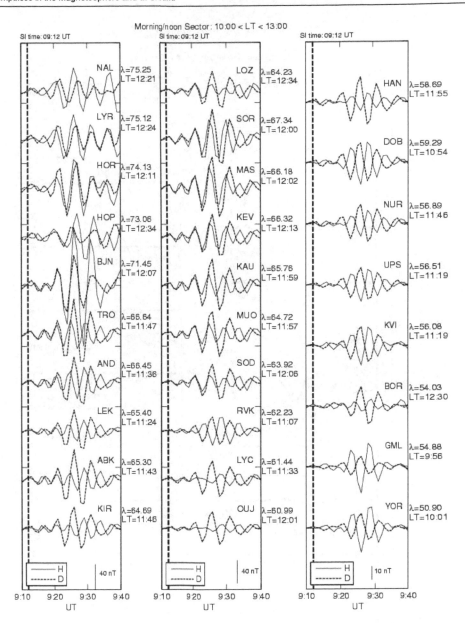

Fig. 7. (after Piersanti et al., 2011) The H (solid) and D (dotted) components of the geomagnetic field at ground stations location in the morning/noon sector (10:00 < LT <13:00) on June 8, 2000 (0910÷0940 UT). The data are filtered at f = 3.3 mHz.

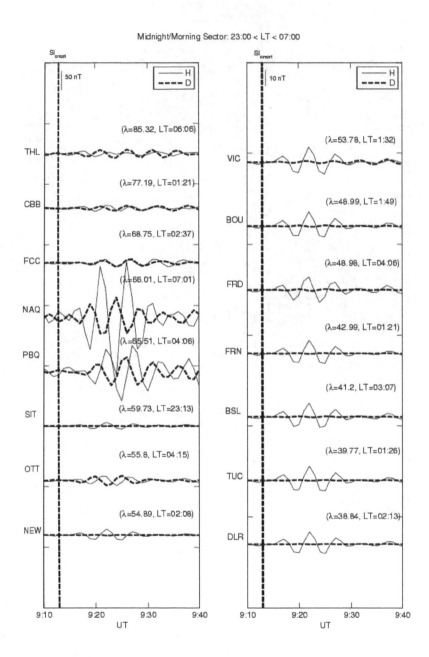

Fig. 8. (after Piersanti et al., 2011) The H (solid) and D (dotted) components of the geomagnetic field at ground stations location in the midnight/morning sector (23:00 < LT < 07:00) on June 8, 2000 (0910÷0940 UT). The data are filtered at f = 3.3 mHz.

5. Summary

The impinging on the Earth's magnetosphere of sudden increase of the SW pressure, related to the arrival of shock waves or other discontinuities, typically causes rapid increases of the magnetospheric and geomagnetic field, called sudden impulses (SI). Such pressure pulses, indeed, compress the magnetosphere, increase the magnetopause and tail currents, and possibly other magnetospheric/ionospheric current systems as well.

On the other hand, since the magnetopause current is mostly enhanced in the dayside sector, while the enhancement of the tail current produces a negative variation of the magnetospheric field, a strong day/night asymmetry might be expected in the SI manifestation: consistently, the experimental observations at geosynchronous orbit reveal an explicit LT modulation, with greater responses at satellite located closer to the noon meridian. The comparison between experimental observations and theoretical models reveals that the positive jumps are basically determined by the changes of the magnetopause current alone, in the dayside and in a large portion of the dark magnetosphere. On the other hand, the occurrence of negligible and negative responses in the nightside reveals a significant role of the tail (and ring) current.

At ground, the SI manifestation is more complex than at geosynchronous orbit since secondary effects like FAC and ionospheric currents contribute significantly in addition to the primary effects of the magnetospheric currents. The ground responses show a clear latitudinal dependence and a LT modulation (with smaller values in the morning and higher values in the postnoon region), a behaviour which can be interpreted as the manifestation of the superimposed ionospheric currents.

SI manifestations are occasionally accompanied by trains of almost monochromatic ULF waves which manifest soon after the main variation and persist for few cycles. In some cases they can be interpreted in terms of global oscillation modes of the whole magnetospheric cavity.

6. Acknowledgements

This research activity is supported by Consorzio Area di Ricerca in Astrogeofisica and Università, L'Aquila.

7. References

Amata, E., V. A. Pilipenko, O. A. Pokhotelov, V. A. Troitskaya, and R. V. Shchepetnov, 1986, PSC-5 Pulsations in Geostationary Orbit, *Geomagn. Aeron.*, 26, (2), 283.

Araki, T., 1977, Global structure of geomagnetic sudden commencements, *Planet. Space Sci.*, 25, 372,.

Araki, T., 1994, A physical model of the geomagnetic sudden commencement, in Solar Wind Sources of Magnetospheric Ultra-Low-Frequency Waves, *Geophys. Mono. Series*, 81, 183.

Araki, T., K. Keika, T.

Araki, T., S. Tsunomura and T. Kikuchi, 2009, Local time variation of the amplitude of geomagnetic sudden commencements (SC) and SC-associated polar cap potential, *Earth Planets Space*, 61, e13.

Araki, T., T. Iyemori, and T. Kamei, 1984, Sudden commencements observed by MAGSAT above the ionosphere, *J. Geomag. Geoelectr.*, 36, 507.

Borodkova, N. L., G. N. Zastenker, M. Riazantseva, and J. D. Richardson, 2005, Large and sharp solar wind dynamic pressure variations as a source of geomagnetic field disturbances in the outer magnetosphere (at the geosynchronous orbit), *Planetary and Space Science*, 53, N.1-3, 25.

Borodkova, N. L., J. B. Liu, Z. H. Huang, G. N. Zastenker, C. Wang, and P. E. Eiges, 2006, Effect of change in large and fast solar wind dynamic pressure on geosynchronous magnetic field, *Chinese Physics*, 15, 2458.

Ferraro, V. C. and H.W. Unthank, 1951, Sudden commencements and sudden impulses in geomagnetism: their diurnal variation in amplitude, *Geophys. Pure Appl.*, 20, 2730.

Francia, P., S. Lepidi, U. Villante, P. Di Giuseppe, 2001, Geomagnetic sudden impulses at low latitude during northward interplanetary magnetic field conditions, *J. Geophys. Res.*, 106, 21231.

Han, D.-S., T. Araki, H.-G. Yang, Z.-T. Chen, T. Iyemori, and P. Stauning, 2007, Comparative study of Geomagnetic Sudden Commencement (SC) between Oersted and ground observations at different local times, *J. Geophys. Res*, 112, A05226, doi:10.1029/2006JA011953.

Kamei, H. Yang, and S. Alex, 2006, Nighttime enhancement of the amplitude of geomagnetic sudden commencement and its dependence on IMF-Bz, *Earth Planets Space*, 58, 45.

Kikuchi, T., S. Tsunomura, K. Hashimoto, and K. Nozaki, 2001, Field aligned current effects on mid-latitude geomagnetic sudden commencements, *J. Geophys. Res.*, 106, 15555.

Kokubun, S., 1983, Characteristics of storm sudden commencement at geostationary orbit, *J. Geophys. Res*, 88, 10025.

Kuwashima, M. and H. Fukunishi, 1985, SSC-associated magnetic variations at the geosynchronous altitude, *J. Atmo. Terr. Phys.*, 47, 451.

Lee, D.-Y., and L. R. Lyons, 2004, Geosynchronous magnetic field response to solar wind dynamic pressure pulse, *J. Geophys. Res*, 109, A04201.

Lühr, H.; Schlegel, K.; Araki, T.; Rother, M.; Förster, M., 2009, Night-time sudden commencements observed by CHAMP and ground-based magnetometers and their relationship to solar wind parameters, *Ann. Geophys.*, 27, 5, 1897.

Matsushita, S., 1962, On geomagnetic sudden commencements, sudden impulses, and storm duration, *J. Geophys. Res.*, 67, 3753.

Nishida, A., 1978, Geomagnetic diagnosis of the magnetosphere, *Phys. Chem. Spa., vol. 9*, Springer-Verlag, New York.

Nishida, A., and J. A. Jacobs, 1962, World-wide changes in the geomagnetic field, *J. Geophys. Res.*, 67, 525.

Patel, V. L., P. J. Jr. Coleman, 1970, Sudden impulses in the magnetosphere observed at synchronous orbit, *J. Geophys. Res.*, 75, 7255.

Piersanti, M., U. Villante, C. Waters, I. Coco, 2012, The June 8, 2000 ULF wave activity: a case study, 2012, in press on *J. Geophys. Res.*.

Russell, C. T., Ginskey, M., Petrinec, S. M., Le, G., 1992, The effect of solar wind dynamic pressure changes on low and mid-latitude magnetic records, *Geophys. Res. Lett.*, 19, 1227.

Russell, C. T., M. Ginskey, 1995, Sudden impulses at subauroral latitudes: response for Northward interplanetary magnetic field, *J. Geophys. Res.*, 100, 23695.

Russell, C. T., M. Ginskey, S. Petrinec, 1994a, Sudden impulses at low-latitude stations: steady state response for Northward interplanetary magnetic fields, *J. Geophys. Res.*, 99, 253.

Russell, C. T., M. Ginskey, S. Petrinec, 1994b, Sudden impulses at low-latitude stations: steady state response for Southward interplanetary magnetic fields, *J. Geophys. Res.*, 99, 13403.

Sanny, J., J. A. Tapia, D. G. Sibeck, and M. B. Moldwin, 2002, Quiet time variability of the geosynchronous magnetic field and its response to the solar wind, *J. Geophys. Res.*, 107(A12), 1443, doi:10.1029/2002JA009448.

Shinbori, A., Y. Tsuji, T. Kikuchi, T. Araki, and S. Watari, 2009a, Magnetic latitude and local time dependence of the amplitude of geomagnetic sudden commencements, *J. Geophys. Res.*, 114, A04217.

Shinbori, A., Y. Tsuji, T. Kikuchi, T. Araki, and S. Watari, 2009b, Magnetic latitude and local time dependence of the amplitude of geomagnetic sudden commencements, *11th IAGA Scientific Assembly*, Sopron.

Siscoe, G. L., V. Formisano, and A. J. Lazarus, 1968, Relation between geomagnetic sudden impulses and solar wind pressure changes: An experimental investigation, *J. Geophys. Res.*, 73, 4869.

Smith, E. J., J. A. Slavin, R. D. Zwickl and S. J. Bame, 1986, Solar Wind-Magnetosphere Coupling and the Distant Magnetotail, *Solar Wind-Magnetosphere Coupling*, 717, eds. Y. Kamide and J. A. Slavin, Terra-Reidel, Tokyo.

Sun, T. R., C. Wang, H. Li, and X. C. Guo, 2011, Nightside geosynchronous magnetic field response to interplanetary shocks: Model results, *J. Geophys. Res.*, 116, A04216, doi:10.1029/2010JA016074.

Tsunomura, S., 1998, Characteristics of geomagnetic sudden commencement observed in middle and low latitudes, *Earth Plan. Spa Phys.*, 50, 755.

Tsyganenko, N. A., 2002a, A model of the near magnetosphere with a dawn-dusk asymmetry. 1. Mathematical structure , *J. Geophys. Res.*, 107, SMP 10-1.

Tsyganenko, N. A., 2002b, A model of the near magnetosphere with a dawn-dusk asymmetry. 2. Parameterization and fitting to observations, *J. Geophys. Res.*, 107, SMP 12-1.

Villante, U. and M. Piersanti, 2008, An analysis of sudden impulses at geosynchronous orbit, *J. Geophys. Res.*, 113, A08213.

Villante, U. and M. Piersanti, 2011, Sudden Impulses at geosynchronous orbit and at ground, *J. Atm. Sol-Terr. Phys.*, 73, 1, 61.

Villante, U., 2007, Ultra Low Frequency Waves in the Magnetosphere, , in: Y. Kamide/A. Chian, *Handbook of the Solar-Terrestrial Environment.* pp. 397, doi: 10.1007/11367758_16, Springer-Verlag Berlin Heidelberg.

Wang, C., et al., 2007, Response of the magnetic field in the geosynchronous orbit to solar dynamic pressure pulses, *J. Geophys. Res.*, 112, A12210.

Wang, C., J. B. Liu, H. Li, Z. H. Huang, J. D. Richardson, and J. R. Kan, 2009, Geospace magnetic field responses to interplanetary shocks, *J. Geophys. Res.*, 114, A05211, doi:10.1029/2008JA013794.

Zhang X. Y., Q. G. Zong, Y. F. Wang, H. Zhang, L. Xie, S. Y. Fu, C. J. Yuan, C. Yue, B. Yang and Z. Y. Pu, 2010, ULF waves excited by negative/positive solar wind dynamic pressure impulses at geosynchronous orbit, *J. Geophys. Res.*, doi:10.1029/2009JA015016.

Ziesolleck, C. W. S., and F. H. Chamalaun, 1993, A Two-Dimensional Array Study of Low-Latitude PC 5 Geomagnetic Pulsations, *J. Geophys. Res.*, 98(A8), 13,703, doi:10.1029/93JA00637.

6

Turbulence in the Magnetosheath and the Problem of Plasma Penetration Inside the Magnetosphere

Elizaveta E. Antonova[1,2], Maria S. Pulinets[1], Maria O. Riazantseva[1,2],
Svetlana S. Znatkova[1], Igor P. Kirpichev[1,2] and Marina V. Stepanova[3]
[1]Skobeltsyn Institute of Nuclear Physics,
Moscow State University, Moscow
[2]Space Research Institute RAS, Moscow
[3]Physics Department, Universidad de Santiago de Chile
[1,2]Russia
[3]Chile

1. Introduction

Chapman & Ferraro (1931) introduced the concept of confinement of the Earth's magnetic field in a cavity carved in the solar plasma flow. The balance between the Earth's magnetic field (more accurately between the magnetic pressure at the boundary of the cavity) and the solar wind dynamic pressure was considered as the condition of the formation of the boundary of the cavity. Chapman-Ferraro model is called a closed magnetosphere. Low energy particles can not penetrate through the boundary of the cavity. Dungey (1961) made the most drastic revision of Chapman-Ferraro's original theory. Dungey envisaged that the connection process, called reconnection, takes place on the dayside magnetopause and that the connected field lines are then transported in the antisolar direction by the solar wind, resulting in the magnetotail. Subsequently, the field lines are reconnected there and then transported back to the dayside magnetosphere. Such process takes place when interplanetary magnetic field (IMF) has the southward direction. The large scale reconnection takes place at high latitudes when IMF has the northward direction. The scheme shown on Fig. 1 demonstrates Dungey's concept of reconnection at the dayside magnetopause when IMF has southword (a) and northward (b) directions. The model of Dungey qualitatively accounts for such phenomena as the inward motion of the dayside magnetopause, equatorward motion of the cusp, expansion of the auroral oval, increase in magnetotail magnetic field strength, and expansion of the magnetotail radius which occur when the IMF turns southward. It can also easily explain the penetration of the plasma of solar wind origin inside the magnetosphere. That is why this concept for a long period was the dominant concept in the physics of the magnetosphere and was widely used for the description of different phenomena including the formation of boundary layers (see, for example, the review Lavraud et al. (2011)). However step by step a number of observations ant theoretical arguments have appeared which give the possibility to throw doubts on the applicability of the scheme shown on Fig. 1 for the real situation.

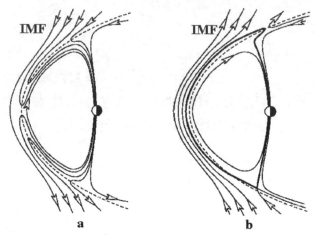

Fig. 1. Sketch illustrating the process of reconnection at the dayside magnetopause when IMF has southward (a) and northward (b) directions

In this paper, we try to summarize arguments demonstrating principal differences of the scheme shown on Fig.1 and the picture, which corresponds to the results of experimental observations. We discuss the process of particle penetration through the magnetopause and try to select arguments demonstrating the formation of the changes of the form of the magnetopause and particle penetration inside the magnetosphere as the results of the change of conditions of pressure balance at the magnetopause. THEMIS mission multisatellite observations available at (http://www.nasa.gov/ mission_pages/themis/) were used for illustration of the main features of magnetic field and plasma observations near the magnetopause. The paper is organized as follows. Section 2 contains the analysis of the properties of turbulence in the magnetosheath. Section 3 is dedicated to the condition of pressure balance at the magnetopause. We discuss the applicability of the frozen in condition for the description of plasma flow in the magnetosheath in Section 4. Section 5 contains conclusions and discussions.

2. Turbulent magnetosheath and magnetic field near the subsolar magnetopause

The magnetopause is formed not as the boundary between the solar wind and the geomagnetic cavity. Magnetosheath plasma and magnetic field come into a contact with magnetopause. The magnetosheath is a region through which mass, energy and momentum are transported from the solar wind into the Earth's magnetosphere. There is a significant number of experimental results showing the high level of plasma turbulence in the magnetosheath (see Luhmann et al. (1986), Sibeck et al. (2000), Zastenker et al. (1999, 2002), Lucek et al. (2001), Němeček et al. (2000a,b; 2002a,b), Shevyrev & Zastenker (2005), Shevyrev et al. (2007), Gutynska et al. (2008), Savin et al. (2008), Rossolenko et al. (2008), Šafránková et al. (2009), Znatkova et al. (2011) and references therein). Fig. 2 shows an example of plasma and magnetic field fluctuations in the magnetosheath measured by Geotail satellite March 2, 1996. It is possible to see that the amplitude of magnetic field fluctuations is much larger than the averaged field and constitutes ~10-20 nTl.

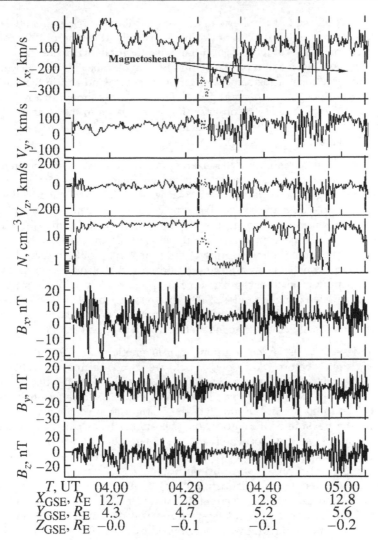

Fig. 2. Results of measurements on the Geotail satellite on March 2, 1996

Zastenker et al. (2002) discussed the origin of magnetosheath variations and showed that a part of these variations is from propagation and/or amplification of solar wind or interplanetary magnetic field (IMF) disturbances, which pass through the bow shock, and a part of these variations originates inside the magnetosheath. Gutynska et al. (2008) present the results of a statistical survey of the magnetosheath magnetic field fluctuations and other parameters using two years of Cluster observations. They have found that the correlation length of the turbulence in the magnetosheath is approximately ~$1R_E$ in the frequency range 0.001– 0.125 Hz and does not depend significantly on the magnetic field or plasma flow direction. When the plasma flow velocity in the magnetosheath is about ~ 200 km/s the distance ~$1R_E$ is traversed by plasma during approximately ~ 30 seconds.

The existence of high level of turbulence in the magnetosheath suggests that the direction of magnetic field near magnetpause can not coincide with the direction of IMF. Šafránková et al. (2009) determine a probability of simultaneous observations of the same sign of the magnetic field B_Z component in the solar wind and magnetosheath. They conclude that the probability of observations of the same B_Z sign in the solar wind and in the magnetosheath is surprisingly very low from a general point of view. It was shown that regardless of the solar cycle phase, the probability to observe the same B_Z sign in the solar wind and in the magnetosheath is close to 0.5 (random coincidence) for IMF $|B_z| < 1$ nT, and it is a rising function of the B_Z value.

Solar wind is the turbulent medium (see Riazantseva et al. (2005, 2007) and references in these works). Therefore, solar wind parameters may change during the propagation to the Earth's orbit from the position of such satellite in the solar wind as ACE and Wind till the orbit of the Earth. That is why to assess the effect of magnetosheath turbulence on the magnetic field parameters changing during the propagation through the magnetosheath to the magnetopause these parameters should be compared directly in front of the shock wave and near the magnetopause. At the same time measuring of the solar wind should be carried out upstream the foreshock region which makes a strong disturbance in the solar wind before the shock front. The opportunity of such a comparison has appeared only with the start of the five-satellite THEMIS mission (Angelopoulos, 2008; Sibeck & Angelopoulos, 2008). One of THEMIS satellites during summer in the north hemisphere performed measurements in the solar wind, while the other occasionally crossed the magnetopause on the dayside.

To obtain the dependences of the component of magnetic field before the magnetopause with magnetic field before the bow shock we used results of THEMIS mission (http://cdaweb.gsfc.nasa.gov/) for the period from June 25 to October 10, 2008. During this period the orbits of spacecrafts deployed by the precession in such a way that their apogees were located close to the Earth-Sun line, i.e. the configuration convenient for studying the interactions on the dayside of the Earth's magnetosphere takes place. The intervals when one of the spacecrafts was localized in the solar wind, and another crossed the magnetopause near the subsolar point were picked out. The events were selected when the deviation of the probe from the x-axis did not exceed 7 R_E. The moment of crossing of the magnetopause was fixed by the distinctive changes in plasma parameters and magnetic field, determined according to the Electrostatic Analyzer ESA (McFadden et al. (2008)) and the Flux Gate Magnetometer FGM (Auster et al. (2008)) on the probe. Parameters of the interplanetary magnetic field (IMF) were determined by FGM. The events in which the solar wind did not suffer significant variations were chosen. The value of the standard deviation of the absolute value of the magnetic field from the average for the selected periods does not exceed 2 nT, the flow velocity was less than 650 km/sec.

The parameters of the magnetic field, measured by one of the spacecraft after crossing the magnetopause, were compared with the IMF parameters, observed by another spacecraft. The following quantities were used as analyzed parameters: the magnitude and the three components of the magnetic field. Mean value and dispersion were calculated for each variable.

The magnetic field parameters near the magnetopause were averaged over periods of 30 and 90 seconds after crossing the magnetopause (what was fixed simultaneously by changes in the parameters of plasma and magnetic field). Values of the magnetic field, averaged over the spin resolution of the probe, equal to 3 s, i.e. field directly close to magnetopause was also analyzed. The solar wind parameters were averaged over a maximum period of 90 s taking into account the time shift of solar wind propagation from the spacecraft performing measurements in the solar wind to the magnetopause. The shift was calculated as the time of the solar wind passing the difference between x-coordinates of the spacecrafts in the approximation of the radial propagation of the solar wind. Solar wind velocity was determined from the data of THEMIS probe located in the solar wind. The solar wind velocity in the magnetosheath was considered as reduced by about two times as a result of thermalization. The magnetosheath thickness was supposed to be approximately ~2 R_E. For each case, the time shift was calculated individually for the specific spacecraft coordinates. Since the errors of the order ten seconds are possible when calculating the time shift, the averaging of values in the solar wind was made for a maximum period of 90 seconds to minimize them. 26 events were analyzed.

Fig. 3–6 show the dependences of the magnetic field parameters near the magnetopause on the solar wind parameters. A set of three curves is given for each parameter. The first distribution is plotted for the instantaneous values (three second averaging) after crossing the magnetopause (panels a), the second – for the averaged over a 30-seconds interval after crossing (panels b), the third – for the averaged over a 90-seconds interval (panels c). The dependencies on the corresponding averaged solar wind parameters are shown. Averaging in the solar wind is realized for a maximum period of 90 seconds (taking into account the time shift of the solar wind propagation to the magnetopause) in order to minimize errors due to deviation of the estimated solar wind delay from the real. For each point, an error calculated as the standard deviation over the averaging periods is also shown. On the charts for the instantaneous values, the errors are shown only for the averaging in the solar wind, because averaging near the magnetopause was not carried out.

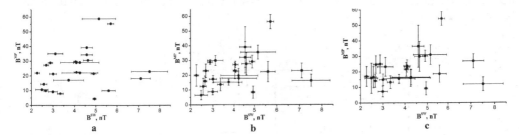

Fig. 3. The dependence of the magnetic field magnitude for the considered set of events a) over 3 seconds after crossing the magnetopause, b) averaged over a period of 30 seconds from the moment of crossing c) averaged over a period of 90 seconds – on the magnitude in the solar wind

The values of the magnetic field magnitude at the magnetopause (see Fig. 3) noticeably trend to increase when increasing magnitude in the solar winds. The form of the distribution remains essentially unchanged when the period of averaging is increased. In accordance with Fig. 4, the X-component of the magnetic field at the magnetopause does not depend on the corresponding value in the solar wind and fluctuates around zero, which is in accordance with the assumption of magnetopause as a tangential discontinuity. As well as for the field magnitude, the increase in averaging interval does not change the form of the

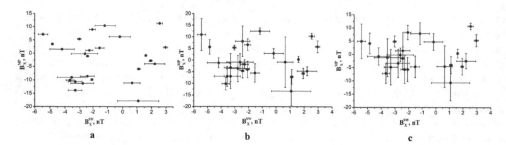

Fig. 4. The dependence of the x-component of the magnetic field for the considered set of events a) over 3 seconds after crossing the magnetopause, b) averaged over a period of 30 seconds from the moment of crossing c) averaged over a period of 90 seconds – on the B_x in the solar wind

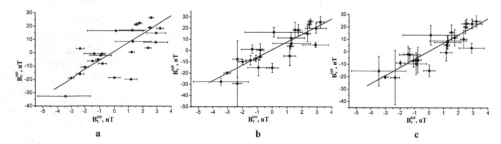

Fig. 5. The dependence of the y-component of the magnetic field for the considered set of events a) over 3 seconds after crossing the magnetopause, b) averaged over a period of 30 seconds from the moment of crossing c) averaged over a period of 90 seconds – on the B_y in the solar wind

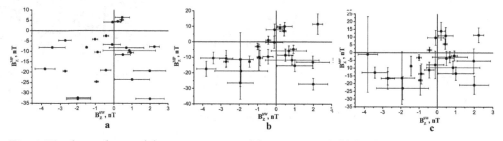

Fig. 6. The dependence of the z-component of the magnetic field for the considered set of events a) over 3 seconds after crossing the magnetopause, b) averaged over a period of 30 seconds from the moment of crossing c) averaged over a period of 90 seconds – on the B_z in the solar wind

distribution of points on the graph. A good linear dependence of the Y-component of the magnetic field at the magnetopause (see Fig. 5) on that in the solar wind is obtained. Therefore B_Y-component at the magnetopause, as it was shown at a stage of the first studies (see Fairfield (1967)), is comparatively well correlated with IMF B_Y. The correlation coefficient increases when increasing averaging period and the errors of the parameters of the approximation and standard deviation is decreased. The results obtained for the Z-component of the magnetic field (see Fig. 6) are of great interest. There is a vague tendency towards an increase in the value of this component with the increasing of corresponding value in the solar wind. However, in at least a quarter of cases (8 out of 26 for the instantaneous values, and 7 out of 26 for the values averaged over a period of 30 seconds and 90 seconds) the sign of the Z-component at the magnetopause changes compared with the sign of B_Z in the solar wind from positive value (in the solar wind) to negative (at the magnetopause), and in a few cases (1 for the instantaneous and 2 averaged over 30 and 90 seconds values) from negative to positive value. Therefore, the correlation of B_Z component near the magnetopause and IMF B_Z is practically absent.

3. Pressure balance at the magnetopause

The problem of pressure balance at the magnetopause continues to be one of the most actual problems of the physics of the magnetosphere as the solution of this problem is deeply connected with the solution of the problem of particle, momentum and energy penetration inside the magnetosphere and the formation of boundary layers. It is necessary to mention that the condition of pressure balance at the magnetopause was not analyzed in connection with Dungey's picture. Sibeck et al. (1991) named Dungey's model the "onion peel" model of magnetic merging. They stressed that "onion peel" model violates pressure balance at the magnetopause and therefore does not lead to a quantitative prediction of the magnetopause location as a function of IMF orientation. Sibeck et al. (1991), Sibeck (1994) verify the pressure balance relationship between the solar wind dynamic pressure and the location of the subsolar magnetopause. It was shown, that the pressure balance between the incident solar wind and the magnetospheric magnetic field determines the location of the dayside magnetopause.

The analysis of the validity of pressure balance at the magnetopause was produced using data of AMPTE/IRM by Phan et al. (1994), Phan & Paschman (1996). Plasma moments are obtained every 4.35 s. The total pressures P_{tot} are obtained by Phan et al. (1994), Phan & Paschman (1996) for cases of small and large magnetic shear across the magnetopause. The perpendicular thermal pressure $P_\perp$ and the magnetic pressure are used to calculate the total pressure. In the low-shear case, both perpendicular plasma pressure and magnetic pressure change significantly across the magnetopause and magnetosheath regions but their sum P_{tot} remains rather constant throughout these regions. In the high-shear case, the magnetosheath magnetic and plasma pressures both remain rather uniform in the entire region within 20 min preceding the magnetopause crossing, so that P_{tot} is also constant. Across the magnetopause, the plasma and magnetic pressures vary significantly, and their sum generally has a small jump across this boundary: a deficiency of P_{tot} on the magnetosheath side and an excess on the low latitude boundary layer (LLBL) side of the magnetopause are often observed. On average, the change of P_{tot} is less than 10%.

Panov et al. (1998) study the pressure balance at high latitude magnetopause using data of CLUSTER mission. Suggestion about thermal pressure isotropy was used during calculation of P_{tot}. It was shown that for most of the analyzed 154 magnetosheath-magnetosphere transitions the pressure balance between the two sides of the transitions was fulfilled within the error bars, the magnetosheath-to-magnetosphere pressure ratio was close to unity, within the range of 0.5 to 2.

THEMIS mission multisatellite observations give the opportunity to clarify the conditions of pressure balance near the subsolar point at the magnetopause having simultaneous observations inside and outside the magnetopause by two satellites quite near to its surface with 3 s resolution. THEMIS Flux Gate Magnetometer (FGM) and the Ion and Electron Electrostatic Analyzer (ESA) data from the THEMIS satellite mission can be used to determine the total plasma pressure with 3 s resolution (McFadden et al., 2008). The magnetometer measures the background magnetic field and its low frequency fluctuations (up to 64 Hz) with amplitudes of 0.01 nT in a range extending over six orders of magnitude (Auster et al., 2008). The electrostatic analyzers measure plasma over the energy range from a few eV up to 30 keV for electrons and 25 keV for ions (McFadden et al., 2008).

An analysis of the pressure balance on the magnetopause near the subsolar point was made for 18 crossings of the magnetopause by the THEMIS project satellites under magneto-quiet conditions by Znatkova et al. (2011). Dynamic and static pressures of plasma are determined, as well as magnetic pressure in the magnetosheath, and magnetic and plasma static pressure inside the magnetosphere. Variations of the total pressure have been studied in the case when one satellite is located inside the magnetosphere and another one stays in the magnetosheath near the magnetopause. It is demonstrated, that for 18 investigated events the condition of pressure balance at the subsolar point is valid on average with an accuracy of 7%, within measurement errors and under applicability of the approximation of anisotropic magnetic hydrodynamics to collisionless plasma of the magnetosheath and magnetosphere.

For this study the event July 22, 2007 of the magnetopause crossing have been selected, using the following criteria: the interplanetary magnetic field has a stable northward orientation and magnetosphere was very quite to exclude the contribution of changes of magnetic fields inside the magnetosphere as a source of magnetopause stress balance destruction. Fig. 7 shows the positions of satellite orbits in GSM coordinate system. Fig. 8a shows interplanetary magnetic field and solar wind dynamic pressure for analyzed event in accordance with Wind data shifted on the time delay of solar wind flow from Wind to the magnetopause (http://cdaweb.gsfc.nasa.gov/). Wind data were used as all THEMIS satellites were near to the magnetopause for the analyzed event. Arrow shows the selected moments of magnetopause crossings. It is possible to see analyzing Fig. 8a that IMF has stably northward orientation. Variations of geomagnetic parameters

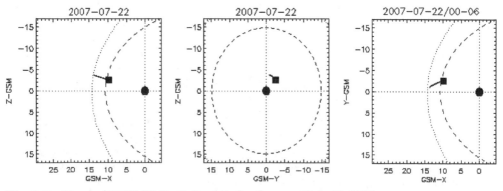

Fig. 7. Positions of THEMIS-C, -D,-E orbits for the event July 22, 2007

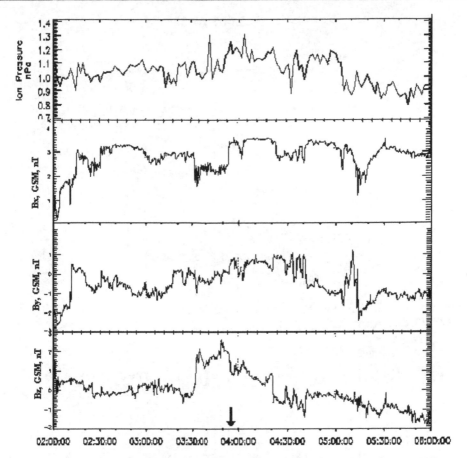

Fig. 8a. Solar wind dynamic pressure and IMF parameters for the event July 22, 2007

(http://swdcdb.kugi.kyoto-u/ac/jp) for the event July 22, 2007 are shown in Fig. 8b. Arrow shows the time for the discussed magnetopause crossings. Dst=-15 hT for the event July 22, 2007. It is possible to see analyzing Fig. 8b that magnetosphere was rather quite having the AE index below 100 nT.

Fig. 9 shows the results of the calculation of dayside magnetic field lines using Tsyganenko-2001 magnetic field model (Tsyganenko, 2002a,b). Stars show the positions of the regions at the magnetic field line where the magnetic field has the minimal value. The values of magnetic field in these regions are also shown. Magnetic field has minimal values at the equator in the inner magnetosphere. These minima are shifted from the equator near the magnetopause. Squares show the position of the equator. Figures near squares show the values of the magnetic field at the equator. It is necessary to mention that Tsyganenko model predicts the existence of closed field lines (dashed lines on Fig. 9) which correspond to the values of the magnetic field at the subsolar point, which can not produce the necessary magnetic pressure for the compensation of the pressure in the magnetosheath.

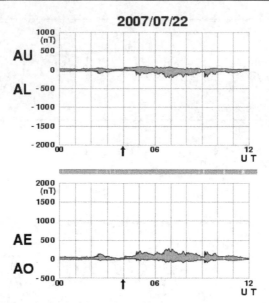

Fig. 8b. AU, AL, AE, AO indexes for the selected time interval in accordance with (http://swdcdb.kugi.kyoto-u/ac/jp)

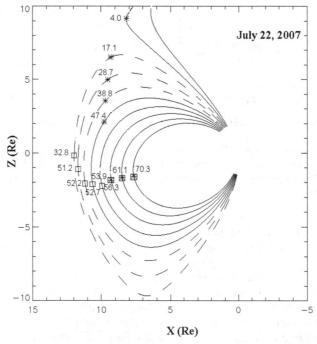

Fig. 9. Forms of the daytime magnetic field line at the Y=0 plane calculated using Tsyganenko-2001 model for solar wind parameters of the event July 22, 2007

Fig. 10 shows the example of magnetic field variations at the moment of THEMIS-D satellite crossing of the magnetopause July 22, 2007. It is possible to see great variations in the amplitude and direction of the magnetic field after the magnetopause crossing, typical for the magnetosheath. Fluctuations of magnetic field ~20 nT are observed. It is necessary to stress that magnetic field has the southward direction after magnetopause crossing in spite of the northward IMF orientating. The solar wind parameters are comparatively stable. Qualitatively the same pictures were observed on other THEMIS satellites. Comparison of the values of the model geomagnetic field (Fig. 9) and the measured geomagnetic field (Fig. 10) shows that amplitudes of magnetic fluctuations in the magnetosheath reach approximately 30 % of the value of magnetic field inside the magnetosphere at the subsolar point. Therefore, fluctuations of magnetic pressure in the magnetosheath can produce ~10% of fluctuations of the total pressure in the magnetosheath. Nevertheless, the same fluctuations at high latitudes, where the magnetic field at the magnetic field line is much smaller than the magnetic field at the subsolar point, the contribution of fluctuations of magnetic pressure is comparable with the magnetic pressure inside the magnetosphere.

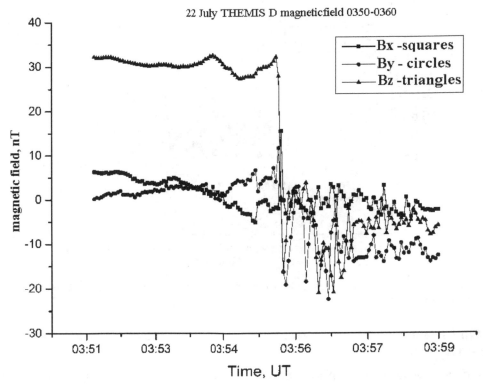

Fig. 10. Variations of THEMIS-D magnetic field July 22, 2007 on the interval 03:51-03:59 UT

The total pressure inside the magnetopause is calculated as a sum of static and magnetic pressure in accordance with the relation $P_{int}=P_{\perp}+(B^2/2\mu_0)$, where μ_0 is the magnetic permeability of vacuum; the integral pressure in the magnetosheath as a sum of dynamic, static and magnetic pressure $P_{int}= n_p\, m_p\, (v_p)^2+ P_{\perp}+(B^2/2\mu_0)$ (see the results of Phan et al.

(1994) and discussion in (Znatkova et al., 2011)), where n_p and v_p are density and velocity (mainly X-component) of protons, m_i is the mass of proton, $P_\perp$ is the component of plasma pressure perpendicular to the magnetic field line. Fig. 11 a,b,c show the total plasma pressure for pairs of satellites for the periods when one satellite was inside the magnetosphere and the other one was inside the magnetosheath. Dashed line shows the results of calculations inside the magnetopause where the MHD analysis of pressure balance cannot be used due to dominance of kinetic effects. As it can be seen in Fig. 11 the total plasma pressure inside and outside magnetosphere is well balanced for all analyzed crossings of the magnetopause just as in the cases discussed by Znatkova et al. (2011). The main difference of the events July 22, 2007 and July 18, 2007 is connected to the total pressure dynamics. The total pressure is nearly constant for the event July 18, 2007 discussed by Znatkova et al. (2011). The total pressure growth is observed for the event July 22, 2007.

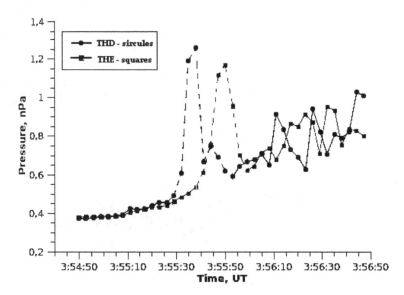

Fig. 11a. Simultaneous measurements of total pressure by the Themis-D and –E satellites for the event July 22, 2007

4. On the applicability of the frozen in condition to the processes in the magnetosheath

The most popular model of plasma flow in the magnetosheath is the gasdynamic model of Spreiter (see Spreiter et al. (1966), Spreiter & Alksne (1969), Spreiter & Stahara (1980)). The validity of this model was demonstrated by Němeček et al. (2000b), Zastenker et al. (2002). It was shown, that Spreiter model describes the parameters of averaged flows of the magnetosheath plasma rather well. MHD models also give the possibility to describe plasma flow parameters. The distribution of magnetic field in the magnetosheath is analyzed in such models using the frozen-in magnetic field approximation. However, it is

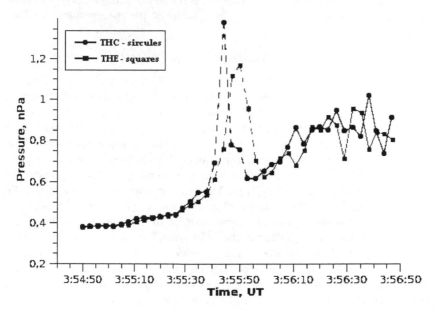

Fig. 11b. Simultaneous measurements of total pressure by the Themis-C and –E satellites for the event July 22, 2007

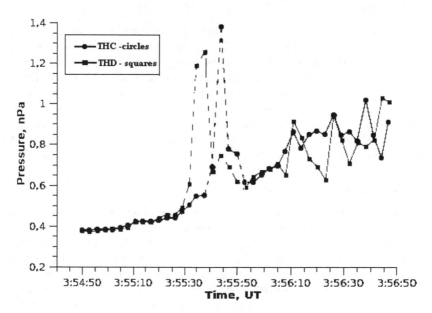

Fig. 11c. Simultaneous measurements of total pressure by the Themis-C and –D satellites for the event July 22, 2007

necessary to stress that frozen-in field condition is the result of the reduction of generalized Ohm's low which have the form in the case of isotropic plasma pressure (see Bittencourt (2004))

$$\mathbf{j} + \tau_{ei}\partial\mathbf{j}/\partial t = \sigma(\mathbf{E} + [\mathbf{VB}] + (\nabla p_e - [\mathbf{jB}])/en)$$

where $\mathbf{j}$ is the current density, τ_{ei} is the electron to ion collision time, $\mathbf{V}$ is the plasma velocity, $\mathbf{E}$ is the electric field, $\mathbf{B}$ is the magnetic field, σ is the conductivity, n is the plasma density and p_e is the electron pressure. $\mathbf{E} + [\mathbf{VB}] = 0$ only if $\sigma \to \infty$ and it is possible to neglect by electron pressure gradient and Hall term. The Reynolds number in the case of Coulomb collisions in the magnetosheath is extremely high (2.5×10^{10} in accordance with (Borovsky & Gary, 2009)). However the existence of high level of turbulence requires the generalized definition of Reynolds number as a ratio of a dissipation time scale τ_{diss} to a convection time scale τ_{conv} for a flow structure ($R = \tau_{diss} / \tau_{conv}$) i.e. the introduction of an effective Reynolds number. Borovsky & Gary (2009) analyzing Landau damping and Bohm diffusion show that the effective Reynolds number is considerably reduced in the magnetosheath in comparison with the Coulomb collision Reynolds number. However, the evaluation of Reynolds number in the magnetosheath gives values >>1. That is why $\sigma \to \infty$ approximation can be used and $\mathbf{E} + [\mathbf{VB}] = ([\mathbf{jB}] - \nabla p_e)/en$. Plasma velocity in the magnetosheath becomes low in comparison with Alfven and sound velocities near the subsolar magnetopause. This means that the equation of motion is reduced to the condition of magnetostatic equilibrium $[\mathbf{jB}] = \nabla(p_e + p_i)$. This means that $\mathbf{E} + [\mathbf{VB}] = \nabla p_i/en$ and frozen-in field condition can not describe the magnetic field in the magnetosheath. The observed plasma pressure anisotropy in the magnetosheath does not lead to principal changes of this conclusion. Results of Sections 2 and 3 also support this conclusion.

The violation of the frozen-in field condition is typical for the plasma sheet of the Earth (see, for example, Borovsky & Bonnell (2001), Troshichev et al. (2002), Stepanova et al. (2009, 2011)). Phan et al. (1994) considered the obtained using AMPTE/IRM results as an evidence for violation of the frozen-in field condition in the magnetosheath as well. Reconnection concept is based on the suggestion of the validity of frozen-in condition, which can be destroyed in a number of points or lines at the magnetopause. However, the results of observations and theoretical analysis show that frozen-in field condition is not applicable for the real magnetosheath and it is necessary to explain the observed penetration of plasma inside the magnetosphere using other suggestions.

5. Conclusions and discussion

Produced analysis demonstrates the real changes of the orientation of magnetic field in the magnetosheath in comparison with the magnetic field before the bow shock including the change of its sign. Poor correlation between the magnetic field in the magnetosheath with the IMF had been noted earlier (see Coleman (2005), Šafránková et al. (2009) and references in these papers). The presented results imply that the poor correlation, even at a relatively long averaging interval of 90 s, comparable with the time of solar wind plasma propagation through magnetosheath, is connected with the magnetosheath turbulization. In this study, due to the limited statistics (limited number of crossings the magnetopause by one of the spacecrafts, when the other was performing measurements upstream the foreshock region) we does not distinguish events with quasiperpendicular and quasiparallel shock waves. In accordance with the results of Shevyrev and Zastenker (2005) one can expect that the average level of fluctuations behind quasiperpendicular and quasiparallel shock waves will differ by about a factor of 2.

Presented results demonstrate the validity of the condition of pressure balance at the magnetopause with comparatively high accuracy and support the results obtained by Phan et al. (1994) using AMPTE/IRM observations. The main difference with Phan et al. (1994) results obtained in this work is connected to a possibility to make simultaneous measurements inside and outside the magnetopause at comparatively small distances using the particle and magnetic field data of the THEMIS satellite mission. Current analysis show that the total pressure is nearly constant during the all satellite crossing of the magnetosheath despite high level of turbulence constantly observed in the magnetosheath. These results show that the total pressure balance is the main condition determining the magnetosheath dynamics and the position of the magnetopause.

Significant number of observations in the magnetosheath, including the THEMIS satellite observations shows the great level of plasma turbulence. Fluctuations of the value and the direction of the magnetic field are constantly present even during periods of comparatively stable orientation of the interplanetary magnetic field. The amplitudes of these fluctuations are comparable with the minimal values of the magnetic field at the dayside magnetic field line (see Rossolenko et al. (2008) and Fig. 9 of this work). These fluctuations create an obstacle for the widely accepted picture shown on Fig. 1 of magnetic reconnection at the magnetopause. Real magnetic field near the magnetopause has different values and orientations at different points of the magnetopause. Which means that ordinarily discussed reconnection picture is nonapplicable for description of the processes near magnetopause. Fig. 12 shows more realistic then the scheme shown on Fig. 1 scheme of magnetic field in the magnetosheath when IMF has northward orientation. Large fluctuations of magnetic field at high latitudes where the magnetic field on the magnetospheric field lines is small (see Fig. 9) create the favorable conditions for field line interconnection and magnetosheath plasma penetration inside the magnetosphere. The events having large amplitudes will be selected as flux transfer events.

IMF, especially its B_z-component, is the major factor controlling the geomagnetic activity. It is usually assumed that this control is performed due to the processes of reconnection of the IMF and the magnetic field on the magnetopause and inside the magnetosphere. Numerous studies of turbulence in the magnetosheath, including the above analysis, give reason to reconsider such suggestion. The high level of magnetic field fluctuations in the magnetosheath, even for the relatively large averaging intervals, indicates that at different points of the magnetopause the magnetic field has different orientations, poorly correlated with the orientation of the IMF. Correspondingly [VB] also has different orientation at different points of the magnetosheath and near the magnetopause. That is why very popular idea on the solar wind electric field penetration inside the magnetosphere has the real obstacles. It is necessary to mention also that the suggestion about large-scale magnetospheric convection generation by boundary layer processes also meets with some obstacles (for example, boundary layers are mapped at the ionosphere near noon in accordance with Newell and Meng (1992)). However, this subject requires analysis that is more careful.

Fig. 12. Sketch illustrating the distribution of magnetic field in the magnetosheath when IMF has northward direction

The ideas about the role of large-scale reconnection processes at the magnetopause and formation of large-scale neutral lines were involved for explaining a relatively good correlation of IMF and large-scale magnetospheric convection. However, it is possible to explain such correlation without postulating the dominant role of reconnection processes at the magnetopause. Let to remind that Sibeck et al. (1991) made a conclusion that "onion peel" model cannot explain the change of magnetopause position under the influence of the interplanetary magnetic field. Sibeck et al. (1991), Tsyganenko & Sibeck (1994), Sibeck (1994) focused on the changes of values of sources of magnetic field inside the magnetosphere under the influence of the IMF and show that such changes can explain the change of magnetopause position under the influence of IMF. Results obtained by Sibeck et al. (1991), Tsyganenko & Sibeck (1994), Sibeck (1994) select the pressure balance at the magnetopause as the main condition determining the dynamics of the magnetopause. It was also shown that the changes of the dayside part of Region 1 currents of Iijima and Potemra under the influence of IMF could produce comparatively large changes in the magnetopause position. Developed by Sibeck et al. (1991) approach is not based on the suggestion of the validity of frozen-in condition, which can be destroyed in a number of points or lines at the magnetopause. The only think, which requires the explanation, is the well-known dependence of the magnetic field inside the magnetosphere on the IMF value and orientation.

Observed distribution of the plasma pressure inside the magnetosphere and of the Region 1 currents of Iijima and Potemra show that Region 1 current generation is a consequence of the existence of azimuthal plasma pressure gradients inside the magnetosphere (see (Iijima et al., 1997; Wing & Newell, 2000; Stepanova et al. 2004; Xing et al., 2009)). These results support the scenario proposed by Antonova & Ganushkina (1997) based on the analysis of the geometry of the high latitude magnetosphere as j=rotB. In this model, the field-aligned currents appear due to divergence of transverse magnetospheric currents. Therefore, the modulation of currents inside the magnetosphere by large-scale IMF can explain the change of magnetopause position under the influence of IMF irrespective to the orientation of the magnetosheath magnetic field near the magnetopause. It was stressed that the external source of the magnetic field in the condition of magnetostatic equilibrium when the gradient of plasma pressure is equal to the Ampere's force (i.e.∇p=[jB])) produces the increase of current in the case of the decrease of B (the addition of field with southward orientation) and the decrease of current in the case of the increase of B (the addition of field with northward orientation) when the plasma pressure gradients change slowly. Therefore, penetration of IMF inside the magnetosphere irrespective of magnetic field fluctuations in the magnetosheath can produce the necessary current modulation. The characteristic time of such modulation is determined by the Alfven travel time (i.e. the time of MHD wave flow from the magnetopause to the distances ~$10R_E$, where plasma pressure gradients which support the Region 1 currents are mainly concentrated according to Xing et al. (2009)). This time is ~2-3 min. Correspondingly, the characteristic time of the change of Region 1 currents and dawn-dusk electric field is of the same order of magnitude. Such estimation is in agreement with the results of radar observations of Ruohoniemi and Greenwald (1998), Ruohoniemi and Baker (1998), Ruohoniemi et al. (2001, 2002) who obtained a small time delay in the response of the high-latitude ionospheric convection to the IMF variations.

Change of the magnetic configuration under the influence of external magnetic field takes place even in the case of vacuum configuration. MHD models of magnetosphere, which do not suggest the validity of the frozen-in condition, clearly demonstrate such influence. Therefore, the solution of the problem does not require the action of reconnection processes at the magnetopause as a reason of IMF influence to the magnetosperic processes. Inversely, such reconnection processes can be a consequence of stress disbalance at the local regions of the magnetopause between the total pressure at the magnetosheth and the mainly magnetic pressure inside the magnetopause. Change of field line topology in such a case has a character of topological reconnection. Large-scale change of magnetic configuration leads to the magnetosheath plasma capture inside the magnetosphere. Such capture was observed, for example, 3 June 2007 by THEMIS satellites and interpreted as an action of double reconnection by Lee at al. (2008). It is interesting to mention that topological reconnection does not require local dissipation (and corresponding plasma heating). High level of magnetosheath turbulence (different values and orientation of magnetic field at different points near the magnetopause) suggests that the discussed reconnection have patchy character. The existence of topological reconnection does not exclude the action of different scenario of classical local reconnection with local destruction of magnetopause current due to development of local instabilities and current dissipation. It only gives the possibility to overcome difficulties related to the presence of high level of turbulence in the magnetosheath.

Suggested scenario helps to overcome difficulties connected with the discussed problem of particle penetration inside the magnetosphere in the conditions of high level of magnetic fluctuations in the magnetosheath. However, it does not exclude the traditionally discussed mechanisms of local destructions of magnetopause current sheet and flux transfer events formation, particle diffusion and development of Kelvin-Helmholts instability at the magnetopause. Future studies will give the possibility to evaluate the importance of each of such mechanisms.

6. Acknowledgments

The authors thank the group of developers of THEMIS mission and the support group of spacecraft data website http://www.nasa.gov/mission_pages/themis/.

We are grateful to V. Angelopoulos and participants in NASA-grant NAS5-02099 for the THEMIS mission realization. In particular: C. W. Carlson and J. P. McFadden for the use of ESA data, K.-H. Glassmeier, U. Auster, and W. Baumjohann for the FGM data, developed under the guidance of Technical University of Braunschweig and with financial support from the German Ministry of Economics and Technology, as well as the German Air and Space Center (DLR) under contract 50 OC 0302.

The work was supported by the grants of President of Russian federation MK-1579.2010.2, Russian Foundation for Basic Research 10-05-00247-a, FONDECYT grant 1110729.

7. References

Angelopoulos, V. (2008). The THEMIS Mission. *Space Sci. Rev.,* Vol. 141, pp. 5–34. doi: 10.1007/s11214-008-9336-1.

Antonova, E. E.; & Ganushkina, N. Yu. (1997). Azimuthal hot plasma pressure gradients and dawn-dusk electric field formation. *J. Atmos. Solar-Terr. Phys.* Vol. 59, No.11, pp.1343-1354.

Auster, H.U.; Glassmeier, K.H., Magnes, W. , Aydogar, O. , Baumjohann, W. et al. (2008). The THEMIS Fluxgate Magnetometer. *Space Sci. Rev.* Vol. 141, pp. 235–264.

Bittencourt, J.A. Fundamentals of Plasma Physics, Springer, 2004, 580 p.

Borovsky, J.E.; & Bonnell, J. (2001). The dc electrical coupling of flow vortices and flow channels in the magnetosphere to the resistive ionosphere. *J. Geophys. Res.* Vol. 106, No A12, pp. 28967-28994.

Borovsky, J. E.; & Gary, S. P. (2009). On shear viscosity and the Reynolds number of magnetohydrodynamic turbulence in collisionless magnetized plasmas: Coulomb collisions, Landau damping, and Bohm diffusion. *Physics of Plasmas.* Vol. 16, 082307. Available from http://pop.aip.org/pop/copyright.jsp

Chapman, S.; & Ferraro, V.C.A. (1931). *Terr. Magn.* Vol. 36, pp. 77.

Coleman, I. J. (2005). A multi-spacecraft survey of magnetic field line draping in the dayside magnetosheath. *Annales Geophysicae.* Vol. 23, pp. 885–900.

Dungey, J.W. (1961). Interplanetary magnetic field and auroral zone. *Phys. Rev. Lett..* Vol. 6, No 1, pp. 47-49.

Fairfield, D. H. (1967). The ordered magnetic field of the magnetosheath. *J. Geophys. Res.* Vol. 72, No 23, pp. 5865-5877.

Gutynska, O.; Šafránková, J. & Němeček, Z. (2008). Correlation length of magnetosheath fluctuations: Cluster statistics. *Annales Geophysicae*. Vol. 26, pp. 2503–2513.

Iijima, T.; Potemra, T.A. & Zanetti, L.J. (1997). Contribution of pressure gradients to the generation of dawnside region 1 and region 2 currents. *J. Geophys. Res.* Vol. 102, No A12, pp. 27069-27081.

Lavraud, B.; Foulton, C., Farrugia, C.J. & Eastwood, J.P. (2011). The magnetopause, its boundary layers and pathways to the magnetotail. *The Dynamic Magnetosphere*, ed. W. Liu and M. Fujimoto, IAGA Special Sopron Book Series, Springer, pp. 3-28.

Li, W.; Raeder, J., Øieroset, M., & Phan, T.D. (2009). Cold dense magnetopause boundary layer under northward IMF: Results from THEMIS and MHD simulations. *J. Geophys. Res.* Vol. 114, A00C15, doi:10.1029/2008JA013497.

Luhmann, J. G.; Russell, C. T., & Elphic, R. C. (1986). Spatial distributions of magnetic field fluctuations in the dayside magnetosheath. *J. Geophys. Res.* Vol. 91, pp. 1711–1715.

Lucek, E. A.; Dunlop, M.W., Horbury, T. S., Balogh, A., Brown, P., Cargill, P., Carr, C., Fornacon, K.-H., Georgescu, E., & Oddy, T. (2001). Cluster magnetic field observations in the magnetosheath: four-point measurements of mirror structures. *Ann. Geophys.* Vol. 19. pp. 1421-1428.

McFadden, J.P.; Carlson, C.W. , Larson, D., Ludlam, M. , Abiad, R. Elliott, B. Turin, P. Marckwordt, M. , & Angelopoulos, V. (2008). The THEMIS ESA plasma instrument and in-flight calibration. *Space Sci. Rev.* Vol. 141, pp. 277-302.

Němeček, Z.; Šafránková, J., Přech, L., Zastenker, G. N., Paularena, K. I., & Kokubun, S. (2000a). Magnetosheath study: INTERBALL observations. *Adv. Space Res.* Vol. 25, pp. 1511-1516.

Němeček, Z.; Šafránková, J., Zastenker, G. N., Pišoft, P., Paularena, K. I., & Richardson, J. D. (2000b). Observations of the radial magnetosheath profile and a comparison with gasdynamic model predictions. *Geophys. Res. Lett.* Vol. 27, pp. 2801–2804.

Němeček, Z.; Šafránková, J., Zastenker, G., Pišoft, P., & Jelínek, K. 2002a. Low-frequency variations of the ion flux in the magnetosheath. *Planet. Space Sci.* Vol. 50, pp. 567–575.

Němeček, Z.; Šafránková, J., Zastenker, G. N., Pišoft, P., & Paularena, K. I. (2002b). Spatial distribution of the magnetosheath ion flux, *Adv. Space Res.* 30(12), 2751–2756.

Newell, P.T. & Meng, C.-I. (1992). Mapping the dayside ionosphere to the magnetosphere according to particle precipitation characteristics. *Geophys. Res. Lett.* Vol. 19, No 6, pp. 609-612.

Panov, E. V.; Büchner, J., Fränz, M., Korth, A., Savin, S. P., Rème, H., & Fornaçon, K.-H. (2008). High-latitude Earth's magnetopause outside the cusp: Cluster observations. *J. Geophys. Res.* Vol. 113, A01220, doi:10.1029/2006JA012123.

Phan, T.-D.; Paschman G., Baumjohanann W., Sckopke N., & Lühr H. (1994). The magnetosheath region adjacent to the dayside magnetopause: AMPTE/IRM observations. *J. Geophys. Res.* Vol. 99, No A1, pp. 121-141.

Phan, T.-D & Paschman G. (1996). Low-latitude dayside magnetopause and boundary layer for high magnetic shear: Structure and motion. *J. Geophys. Res.* Vol. 101, No A4, pp. 7801-7815.

Riazantseva, M.O.; Zastenker G.N., Richardson J.D., & Eiges P.E. (2005). Sharp boundaries of small- and middle-scale solar wind structures. *J. Geophys. Res.* Vol. 110, A12110, doi:10.1029/2005JA011307.

Riazantseva, M.O.; Khabarova O.V., Zastenker G.N., & Richardson J.D. (2007). Sharp boundaries of solar wind plasma structures and their relationship to solar wind turbulence. *Adv. Space Res.* Vol. 40, pp. 1802–1806.

Rossolenko, S. S.; Antonova, E. E. , Yermolaev, Yu. I., Verigin, M. I., Kirpichev, I. P., & Borodkova, N. L. (2008). Turbulent fluctuations of plasma and magnetic field parameters in the magnetosheath and the low-latitude boundary layer formation: Multisatellite observations on March 2, 1996. *Cosmic Research.* Vol. 46, No 5, pp. 373–382.

Ruohoniemi, J. M.; & Greenwald, R. A. (1998). The response of high latitude convection to a sudden southward IMF turning. *Geophys. Res. Lett.* Vol. 25, No 15, pp. 2913-2916.

Ruohoniemi, J.M.; & Baker, K.B. (1998) Large-scale imaging of high-latitude convection with Super Dual Auroral Radar Network HF radar observations, *J. Geophys. Res.* 103(A9), 20797-20811.

Ruohoniemi, J.M.; Barnes, J.M., Greenwald, R.A., Shepherd, S.G., & Bristow, W.A. (2001). The response of high latitude ionosphere to the CME event of April 6, 2000: a practical demonstration of space weather now casting with the SuperDARN HF radars. *J. Geophys. Res.* Vol. 106, No A12, pp. 30085-30097.

Ruohoniemi, J. M.; Shepherd, S.G., & Greenwald, R.A. (2002). The response of the high-latitude ionosphere to IMF variations. *J. Atmosph. and Solar-Terr. Physics.* Vol. 64, No. 2, pp. 159-171.

Šafránková, J.; Hayosh, M., Gutynska, O., Němeček, Z., Přech, L. (2009). Reliability of prediction of the magnetosheath B_Z component from interplanetary magnetic field observations. *J. Geophys. Res.* Vol. 114, A12213, doi:10.1029/2009JA014552.

Savin, S.; Amata, E., Zelenyi, L., Budaev, V. et al. (2008). High energy jets in the Earth's magnetosheath: Implications for plasma dynamics and anomalous transport. *JETP Letters.* Vol. 87, No 11, pp. 593–599.

Shevyrev, N. N.; & Zastenker, G. N. (2005). Some features of the plasma flow in the magnetosheath behind quasi-parallel and quasi-perpendicular bow shocks. *Planet. Space Sci.* Vol. 53, pp. 95–102.

Shevyrev, N. N., Zastenker, G. N., & Du, J. (2007). Statistics of low-frequency variations in solar wind, foreshock and magnetosheath: INTIERBALL-1 and CLUSTER data. *Planet. Space Sci.* Vol. 55, No 15, pp. 2330–2335.

Sibeck, D.G.; Lopez, R.E., & Roelof, E.C. (1991). Solar wind control of the magnetopause shape, location and motion. *J. Geophys. Res.* Vol. 96, No A4, pp. 5489-5495.

Sibeck, D.G. (1994). Signatures of flux erosion from the dayside magnetosphere. *J. Geophys. Res.* Vol. 99, No A5, pp. 8513-8529.

Sibeck, D.G.; & Angelopoulos, V. (2008). THEMIS science objectives and mission phases,. *Space Sci. Rev..* Vol. 141, pp. 35–59, doi: 10.1007/s11214-008-9393-5.

Sibeck, D.G.; Phan T.-D., Lin R.P. et al. (2000). A survey of MHD waves in the magnetosheath: International Solar Terrestial Program observations. *J. Geophys. Res.* Vol. 105, No A1, pp. 129–138.

Spreiter, J.R.; Summers, A.L., & Alksne, A.Y. (1966). Hydromagnetic flow around the magnetosphere. *Planet. Space. Sci.*. Vol. 14, No 3, pp. 223-253.

Spreiter, J.R.; & Alksne, A. Y. (1969). Plasma flow around the magnetosphere. *Rev. Geophys.*. Vol. 7, pp. 11-50.

Spreiter, J.R.; & Stahara, S.S. (1980). A new predictive model for determining solar wind-terrestrial planet interaction,. *J. Geophys. Res.* Vol. 85, No 12, pp. 6769- 6777.

Stepanova, M.V.; Antonova, E.E., Bosqued, J.M., Kovrazhkin, R. (2004). Azimuthal plasma pressure reconstructed by using the Aureol-3 satellite date during quiet geomagnetic conditions. *Adv. Space Res.* Vol. 33, No 5, pp. 737–741, doi:10.1016/S0273-1177(03)00641-0.

Stepanova M.; Antonova, E. E., Paredes-Davis, D., Ovchinnikov, I. L. & Yermolaev, Y. I. (2009). Spatial variation of eddy-diffusion coefficients in the turbulent plasma sheet during substorms. *Ann. Geophys.* Vol. 27, pp. 1407–1411.

Stepanova, M.; V. Pinto, J. A. Valdivia, & E. E. Antonova (2011) Spatial distribution of the eddy diffusion coefficients in the plasma sheet during quiet time and substorms from THEMIS satellite data. *J. Geophys. Res.*, Vol. 116, A00I24, doi:10.1029/2010JA015887.

Troshichev, O.A.; Antonova, E.E., & Kamide, Y. (2002). Inconsistence of magnetic field and plasma velocity variations in the distant plasma sheet: violation of the "frozen-in" criterion? *Adv. Space Res.* Vol. 30, No. 12, pp. 2683-2687.

Tsyganenko, N.A.; & Sibeck, D.G. (1994). Concerning flux erosion from the dayside magnetosphere. *J. Geophys. Res.* Vol. 99, No A7, pp. 13425-13436.

Tsyganenko, N.A. (2002a). A model of the near magnetosphere with a dawn-dusk asymmetry: 1. Mathematical structure. *J. Geophys. Res.* Vol. 107, No A8, doi: 10.1029/2001JA000219.

Tsyganenko, N.A. (2002b). A model of the near magnetosphere with a dawn-dusk asymmetry: 2. Parameterization and fitting to observations. *J. Geophys. Res.* Vol. 107, No A8, doi: 10.1029/2001JA000220.

Zastenker, G. N.; Nozdrachev, M. N., Němeček, Z., Šafránková, J., Přech, L., Paularena, K. I., Lazarus, A. J., Lepping, R. P., & Mukai, T. (1999). Plasma and magnetic field variations in the magnetosheath: Interball-1 and ISTP spacecraft observations, in: Interball in the ISTP Program, edited by: Sibeck, D. G. and Kudela, K., *NATO Science Series*. Vol. 537, pp. 277–294.

Zastenker, G. N.; Nozdrachev, M. N., Němeček, Z., Šafránková, J., Paularena, K. L., Richardson, J. D., Lepping, R. P., & Mukai, T. (2002). Multispacecraft measurements of plasma and magnetic field variations in the magnetosheath: Comparison with Spreiter models and motion of the structures. *Planet. Space Sci.* Vol. 50, pp. 601–612.

Znatkova, S. S.; Antonova, E. E., Zastenker, G. N. &Kirpichev, I. P. (2011). Pressure balance on the magnetopause near the subsolar point according to observational data of the THEMIS project satellites. *Cosmic Research*. Vol. 49, No. 1, pp. 3–20.

Wing, S.; & Newell, P.T. (2000). Quite time plasma sheet ion pressure contribution to Birkeland currents. *J. Geophys. Res.* Vol. 105, No A4, pp. 7793-7802.

Xing, X.; Lyons, L.R, Angelopoulos, V., Larson, D., McFadden, J., Carlson, C., Runov, A., &
    Auster, U. (2009). Azimuthal plasma pressure gradient in quiet time plasma sheet.
    Geophys. Res. Lett. Vol. 36, No 14, L14105. doi:10.1029/2009GL038881.

7

Ground-Based Monitoring
of the Solar Wind Geoefficiency

Oleg Troshichev

Arctic and Antarctic Research Institute

Russia

1. Introduction

The Earth's magnetosphere is a result of the solar wind impact on the dipole-like geomagnetic field. The form and size of geomagnetosphere are basically determined by the solar wind dynamic pressure, whereas unsteady processes within the magnetosphere affecting the human activity under name of magnetic storms and substorms are due to variations of the magnetic field transported by the solar wind plasma. This field, known as interplanetary magnetic field (IMF), is governed by solar activity and can reverse its direction and vary in amplitude a ten times. As soon as the space-borne measurements of solar wind parameters started, it became clear that southward IMF (opposite in direction to the Earth magnetic field) is one of the most geoeffective solar wind characteristics. The perfect relationship was found between southward IMF (B_{ZS}) and such indicators of magnetic activity as a planetary Kp index (Fairfield & Cahill, 1966; Wilcox et al., 1967; Rostoker & Fälthammar, 1967; Baliff et al., 1967), AE index of magnetic activity in the auroral zone (Pudovkin et al., 1970; Arnoldy, 1971; Foster et al., 1971; Kokubun, 1972; Meng et al., 1973), magnetic storm Dst index (Hirshberg & Colburn, 1969; Kokubun, 1972; Kane, 1974; Russel et al., 1974). Since the correlation of magnetic activity with the solar wind fluctuations distinctly increased when the product of the solar wind speed and southward IMF was taken into account (Rostoker & Fälthammar, 1967; Garrett et al., 1974; Murayama & Hakamada, 1975), the conclusion was made that the interplanetary electric field $E=vxB_{ZS}$ plays a crucial part in the solar wind–magnetosphere coupling (Rostoker & Fälthammar, 1967).

Various combinations of solar wind parameters (basically, the interplanetary electric field and the solar wind density and speed) were repeatedly examined to establish the best function for description of the solar wind-magnetosphere coupling. The most well-known functions are an parameter $\varepsilon=l_0^2 vB^2 sin^4(\theta_c/2)$, firstly presented by Perreault & Akasofu (1978), electric field $E_{KL} = vB_T sin^2(\theta_c/2)$ introduced in practice by Kan & Lee (1979), and rectifying function $E_Y= vB_{ZS}$ (Reiff & Luhmann, 1986), where v is a velocity of the solar wind, B is IMF intensity, B_Y, B_Z and B_{ZS} are azimuthal, vertical and southward IMF components, B_T is transverse IMF component $B_T= \{(B_Y)^2+(B_Z)^2\}^{1/2}$, and θ_c is an angle between B_T component and the geomagnetic Z-axis. A precise formula for the solar wind-magnetosphere coupling function has not yet been agreed so far, and other functions, in number of more than ten, are also used in practice.

In attempts to derive an universal solar wind–magnetosphere coupling function, Newell et al. (2007, 2008) investigated behavior of 10 variables, characterizing the magnetosphere state, in relation to different coupling functions. The comprehensive investigations did not reveal a unique coupling function applicable for any circumstances and conditions, but it was noted (Newell et al., 2008), that the unique coupling function, if it exists, must involve the solar wind velocity v to the first (or a little higher) power, transverse IMF component B_T to the first (or a little lower) power, and sine of IMF clock angle θ_c to the second (or more) power. It is easy to see that coupling function E_{KL} (Kan & Lee, 1979) is well consistent with these requirements.

Coupling functions are used to characterize the solar wind geoefficiency. As this takes place, the solar wind parameters are measured outside the magnetosphere, at present on board ACE spacecraft positioned in Lagrange point (L1) at a distance of ~ 1.5 millions km from Earth. As a consequence, an actual value of the solar wind parameters at magnetopause can essentially differ from those monitored on board ACE spacecraft, even if they are time-shifted to the magnetosphere. Besides, a very high level of magnetic field turbulence is typical of region between the bow shock and magnetopause with incorporation of nonlinear processes within the boundary magnetosphere (Rossolenko et al., 2009), and it is unlikely to wait that changes in the solar wind parameters are converted in their true shape into the magnetosphere processes, while transmitting a signal through the highly turbulent region. Hence it is very desirable to monitor the solar wind energy that entered into the magnetosphere. It is suggested to use for this purpose a ground-based PC index put forward by Troshichev et al. (1988) as a index of magnetic activity in the polar cap.

2. Physical backgrounds and method for the PC derivation

Distribution and intensity of magnetic activity in the Earth's polar caps is determined by orientation and power of the interplanetary magnetic field, particular types of the polar cap magnetic disturbances being related to the IMF southward B_{ZS} (Nishida, 1968) , azimuthal B_Y (Svalgaard, 1968; Mansurov, 1969) and northward B_{ZN} (Maezawa, 1976; Kuznetsov & Troshichev, 1977) components. Distribution of magnetic disturbances on the ground level is commonly described by systems of equivalent currents being hypothetic currents, providing the observed magnetic effect on the ground surface. Figure 1 demonstrates DP2, DP3 and DP4 current systems derived for polar cap disturbances associated with action of southward, northward and azimuthal IMF components, respectively (Kuznetsov & Troshichev, 1977; Troshichev & Tsyganenko, 1978). In addition, the DP0 disturbances have been separated (Troshichev & Tsyganenko, 1978) which are observed irrespective of the IMF, but well correlate with the solar wind velocity v in the second power (Sergeev & Kuznetsov, 1981) that makes it possible to associate them with the solar wind dynamic pressure. DP2 and DP0 current systems are terminated by the latitudes of $\Phi=50\text{-}60°$ (Troshichev, 1975) and focuses in these current vortices are located just right where the intense magnetospheric field-aligned currents are regularly observed.

Measurements on board spacecrafts OGO-4 (Zmuda & Armstrong, 1974) and TRIAD (Iijima & Potemra, 1976) showed that field-aligned currents (FAC) are distributed in two regions aligned with the auroral oval. Region 1 FAC system consists of a layer of field-aligned currents on the poleward boundary of the auroral oval, with currents flowing into the ionosphere in the morning sector and flowing out of the ionosphere in the evening sector.

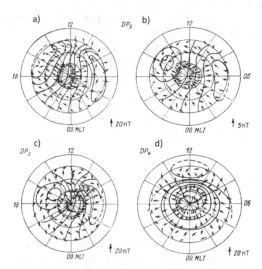

Fig. 1. Current systems of DP2, DP3 and DP4 disturbances generated by variations of IMF components: (a) southward $B_{ZS}=-1nT$, (b) southward $B_{ZS}=-0.25nT$, (c) northward B_{ZN}, (d) azimuthal B_Y (Troshichev and Tsyganenko, 1978).

Region 2 FAC system with the oppositely directed field-aligned currents is positioned on the equatorward boundary of the oval. The currents in Region 1 demonstrate a strong dependency on southward IMF B_{ZS} (Langel, 1975; Mc Diarmid et al., 1977; Iijima & Potemra, 1982) or the interplanetary electric field E_Y (Bythrow & Potemra, 1983). The Region 1 FAC system is observed permanently, even during very quiet conditions, whereas Region 2 system becomes available when magnetic disturbances in the auroral zone are developed (Iijima & Potemra, 1978). Regions 1 and 2 FAC systems are mapped into the equatorial plane of magnetosphere in all magnetosphere models (Antonova et al., 2006). Observations of ion fluxes and magnetic fields made on board DMSP F7 satellite (Iijima et al., 1997), plasma sheet temperature, density, and pressure data inferred from DMSP F8, F9, F10, and F11 satellite measurements at the ionospheric altitudes (Wing and Newell, 2000), measurements of plasma pressure gradients on board the THEMIS satellites (Xing et al., 2009) lead to conclusion that the field-aligned current systems are constantly driven by the pressure-gradient forces generated within the closed magnetosphere while it's coupling with the varying solar wind. The azimuthal pressure gradients required to support the Region 1 field-aligned currents have been derived in studies (Stepanova et al., 2004, 2006; Antonova et al., 2011)

The field-aligned currents patterns typical of low and high magnetic activity have been applied to calculate the systems of electric fields and currents in the polar ionosphere (Nisbet et al., 1978; Gizler et al., 1979; Troshichev et al., 1979). The calculated equivalent current systems turned out to be identical to the experimental systems derived from magnetic disturbances observed in the polar caps. The conclusion was made (Troshichev, 1982) that magnetic activity in polar caps is related to the field-aligned currents responding to changes in the solar wind, and the DP2 disturbances can be taken as an indicator of the geoeffective solar wind impacting on the magnetosphere.

Examination of statistical relationships between the DP2 magnetic disturbances and different interplanetary quantities (Troshichev and Andrezen, 1985) gave the best result for coupling function E_{KL}: the disturbance values δF at the near-pole stations Thule (Greenland) and Vostok (Antarctica) turned out to be linearly linked with E_{KL} value. It means that E_{KL} value can be estimated if the scaling coefficients between quantities E_{KL} and δF is established. Dependence of δF value on ionospheric conductivity is easily taken into account under common conditions, when the polar cap ionosphere is regulated by the solar UV irradiation. These are the physical backgrounds determining a method for the PC index derivation (Troshichev et al., 1988, 2006).

The statistically justified regression coefficients a and β, which determine the relationship between the coupling function E_{KL} and vector of DP2 magnetic disturbance δF at stations Thule and Vostok, are derived at first:

$$\delta F = \alpha\, E_{KL} + \beta \tag{1}$$

These coefficients are calculated for any UT moment of each day of the year since orientation of DP2 disturbances at stations is dependent on local time and slightly changed from summer to winter. To determine the statistically justified orientation of the DP2 disturbances vector (i.e. angle ϕ between the disturbance vector and the dawn-dusk meridian), the correlation between δF and E_{KL} is calculated for all angles ϕ in the range of $\pm90°$ from the dawn-dusk orientation on the basis of large set of data (some years). The angle ϕ ensuring the best correlation between values δF and E_{KL} is fixed and the appropriate regression coefficients a and β are calculated. Just these parameters ϕ, a and β are used in further calculations.

The values E_{KL} were calculated from measurements of solar wind parameters in space, shifted to the sub-solar point (12 RE) using the actual solar wind velocity. Then a time delay $\Delta T \sim 20$ min is required for an E_{KL} signal to be transferred from the bow shock position to the polar cap. To take into account the effect of ionospheric conductivity variations and its changes in response to solar activity, the δF values were estimated in reference to quiet daily curve (QDC) with allowance for QDC change from day to day (Janzhura & Troshichev, 2008). Usage of a proper QDC, as a level of reference for δF values, ensures invariance of the parameters α, β and ϕ determining relationship between δF and E_{KL} irrespective of solar activity (Troshichev et al., 2011a). As an evidence, **Figure 2** shows distribution of α, β and ϕ parameter derived for the Vostok station for epochs of solar maximum (Troshichev et al. 2006) and solar minimum (Troshichev et al., 2011a), and for entire cycle of solar activity (Troshichev et al., 2007, Troshichev & Janzhura, 2009). One can see that patterns for ϕ, α and β parameters derived independently for different epochs are totally consistent (if make allowance for some difference in their scales). It means that once derived parameters of α, β and ϕ can be regarded as valid for ever provided the appropriate QDCs are used (additional substantiations can be found in (Troshichev et al., 2011a)).

The parameters a, β and ϕ established for each UT moment of each day of the year are further used for calculation of PC index of any given time

$$PC = \xi\,(\delta F - \beta)/\,\alpha \tag{2}$$

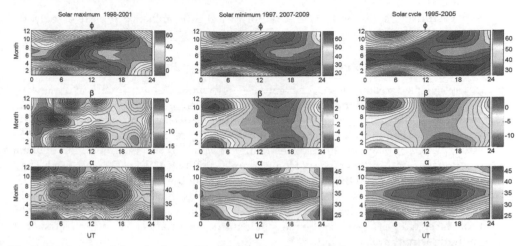

Fig. 2. Parameters ϕ, β and a for the Vostok station, derived independently for epochs of solar maximum (1998–2001), solar minimum (1997, 2007–2008), and the full cycle of solar activity (1995–2005); the axis of abscises being for UT and axis of ordinates being for a month.

The normalization coefficients α and β derived independently for the Thule and Vostok stations eliminate the diurnal and seasonal changes in response of the PC index to changes in E_{KL} field in the summer and winter polar caps. Dimensionality of the scale coefficient ξ is taken equal to 1 for convenience of comparison of PC and E_{KL} values. As an example, **Figure 3** shows the run of the calculated PCN (blue) and PCS (red) indices in the northern and southern hemispheres in 1998-2001. One can see a remarkable agreement in behavior of the positive PCS and PCN indices (which are related to DP2 disturbances) irrespective of the season, the largest value of PC about +20 mV/m being reached synchronously at both stations. Asymmetry is seen for negative PC index which describes the effect of DP3

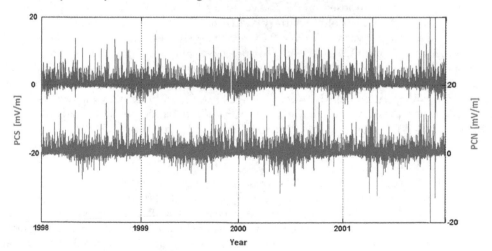

Fig. 3. Run of PCN and PCS indices in 1998–2001 (Troshichev et al., 2006).

magnetic disturbances responding to the northward IMF influence typical of only the summer polar cap.

3. PC index response to changes in solar wind dynamic pressure and E_{KL}

Statistically justified relationships between PC index and changes in the interplanetary electric field E_{KL} and solar wind dynamic pressure P_{SW} were examined in (Troshichev et al., 2007) on the example of the interplanetary shocks, which are commonly accompanied by the clearly defined and significant solar wind dynamic pressure pulses and strong oscillations of the interplanetary magnetic field. To estimate the value E_{KL} and the solar wind dynamic pressure P_{SW} the 5-min averaged data from the ACE satellite for 1998-2002 were used. These parameters were reduced to the magnetopause, the time of the signal passage from the ACE location to the magnetosphere been taken into account with allowance for the real solar wind speed for each particular event. Allowance for additional delay time $\tau_D \sim 20$ minutes typical of the signal passage from the magnetopause to the polar cap and its transformation into magnetic activity was not made in this case.

Only interplanetary shocks with sudden pressure pulses $\Delta P_{SW} > 4$ nPa starting against the background of the steady quiet pressure level lasting no less than 6 hours were examined in the analysis (N=62). The moment of maximum derivative dP/dt was identified as a pressure pulse onset. Just this moment is taken as a key ("zero") date for the epoch superposition method, other characteristics – E_{KL}, PC, and real P_{SW} – being related to the key date. PC indices were classified as summer and winter ones ($PCsummer$ and $PCwinter$), instead of PCN and PCS indices. To separate the overlapping effects of the electric field and dynamic pressure, the behavior of averaged characteristics P_{SW}, E_{KL} and PC was examined under different restrictions imposed in turn on (1) the coupling function E_{KL}, (2) the magnitude of the pressure P_{SW} after the jump, and (3) the rate of the pressure increase (dP_{SW}/dt), the other two quantities being successively kept invariant at the same time.

Analysis of the relationships between the averaged E_{KL} and PC quantities under conditions of varying restrictions imposed on the E_{KL} value ($1 > E_{KL} > 0$ mV/m, $3 > E_{KL} > 1$ mV/m, $E_{KL} > 3$mV/m) with the practically arbitrary values of the pressure jump (pressure gradient $\Delta P_{SW} > 2$ nPa and derivative $dP/dt > 0.04$ nPa/min) showed that the PC index starts to growth within few minutes after the pressure jump, almost simultaneously with the E_{KL} increase. Nevertheless, the maximal magnetic activity in the polar caps is reached about 15 – 30 minutes after the E_{KL} maximums, the corresponding average E_{KL} and PC values being almost identical. Relationships between averaged P_{SW}, E_{KL} and PC under varying restrictions imposed on the pressure magnitude suggest that the dynamic pressure gradient, not the pressure level, affects the PC index. Indeed, the PC index rose just after the pressure jump and descended about 1 hour later irrespective of the persistent high level of the dynamic pressure.

Figure 4 demonstrates relationship between the averaged P_{SW}, E_{KL} and PC values under varying restrictions imposed on the pressure growth rate (dP/dt): (a) $0.2 > (dP/dt) > 0.1$ nPa/min, (b) $0.3 > (dP/dt) > 0.2$ nPa/min and (c) $(dP/dt) > 0.3$ nPa/min, the values of E_{SW} and ΔP_{SW} being arbitrary. One can see that the electric field E_{KL} increases when the dynamic pressure growth rate arises: the sharper is front of the pressure enhancement, the larger are the changes of E_{KL} on the front. Average PC indices and E_{KL} start to increase some minutes

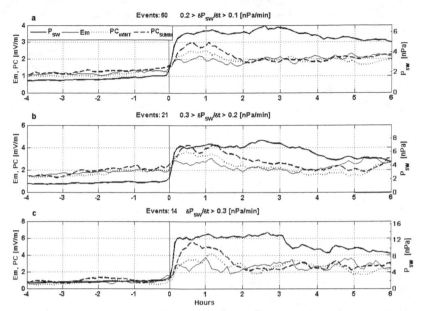

Fig. 4. Relationship between averaged P_{SW}, E_{KL} and PC quantities under varying restrictions imposed on of the pressure growth rate (dP/dt): (a) $0.2 > (dP/dt) > 0.1$ nPa/min, (b) $0.3 > (dP/dt) > 0.2$ nPa/min and (c) $(dP/dt) > 0.3$ nPa/min, the values of E_{KL} and ΔP_{SW} being arbitrary (Troshichev et al., 2007) E_{KL} quantity is denoted here as Em.

after the pressure jump, and the PC indices maximums follow the electric field maximums with a common delay time of ~ 15 – 30 minutes. The PC index remains increased only 1–1.5 hours after a sharp pressure increase, and then the polar cap magnetic activity quickly decays in agreement with E_{KL} behavior.

The excess of the PC index over the appropriate E_{KL} value is typical of the pressure jump conditions: $PCsum$ ~ 3 mV/m for E_{KL} ~ 2 mV/m and $0.2 > (dP/dt) > 0.1$ nPa/min (Fig. 4a); $PCsum$ ~ 4.2 mV/m for E_{KL} ~ 2.5 – 2.8 mV/m and $0.3 > (dP/dt) > 0.2$ nPa/min (Fig. 4b); $PCsum$ ~ 5.6 mV/m for E_{KL} ~ 2.5 – 3 mV/m and $(dP/dt) > 0.3$ nPa/min (Fig. 4c). The discrepancies between the values of E_{KL} and PC indices turn out to be proportional to the corresponding values of ΔP_{SW} and can be assigned just to the influence in the solar wind dynamic pressure. It makes it possible to conclude that an effect of pressure gradient ~1 nPa is approximately equivalent to effect of $E_{KL} \approx 0.33$ mV/m.

A decisive argument in favor of the pressure gradients influence on polar cap activity could be provided with events when the pressure jumps are not accompanied by electric field changes. Unfortunately, it is not feasible to find sharp pressure increases, which are inconsistent with no any variations of the interplanetary electric field. However, there is a number of pressure decreases divorced from the electric field. Results of the epoch superposition for 94 events with negative pressure gradients ($-0.1 > dP_{SW}/dt$ nPa/min) are presented in **Figure 5a**, where the moment of pressure sudden decrease is taken as a "zero moment". One can see that both summer and winter PC indices started to decrease right after "zero moment" although the mean electric field remained at level of ~ 2.5 mV/m

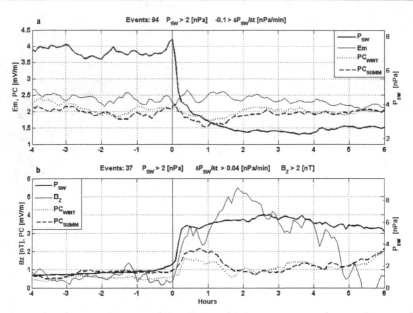

Fig. 5. Relationship between averaged P_{SW}, E_{KL} and PC quantities under conditions of (a) negative pressure gradient ($-0.1 > dP_{SW}/dt$ nPa/min), and (b) northward IMF (Troshichev *et al.*, 2007) E_{KL} quantity is denoted here as Em.

before and after the key date. The average decay of the polar cap magnetic activity lasted about 1.5 hours, and the effect was maximal ($\sim$ 0.5 mV/m) in the summer polar cap. A pure effect of the pressure negative jump in these events is estimated as 1 nPa $\approx$ 0.25 mV/m. The fact that the PC index decreases in response to negative pressure gradients suggests that the pressure effect in the PC index is reversible and acts under conditions of the pressure rise and fall.

Figure 5b shows relationship between averaged P_{SW} and E_{KL} under conditions of northward IMF. Since the coupling function E_{KL} reduces to zero when the IMF is northward ($B_Z > 0$), the run of the IMF B_Z component is shown in Figure. One can see that the mean PC index increases in response to positive pressure pulses, the ratio 1 nPa $\approx$ 0.4 mV/m is valid in this case. Thus, the solar wind pressure growth rate (i.e. jump power $\Delta P_{SW}/\Delta t$) proves, after E_{KL}, to be the second most important factor for the PC index increase: influence of the dynamic pressure gradient $\Delta P_{SW} = 1$ nPa on the polar cap magnetic activity is roughly equivalent to effect of the coupling function $\Delta E_{KL} = 0.33$ mV/m.

The PC index rises in response to the positive dynamic pressure pulses irrespective of IMF polarity (southward or northward). It means that mechanism of the pressure gradients influence on polar magnetic activity is not related to IMF whose effects in polar ionosphere are strongly controlled by the IMF orientation. The PC indices in the summer and winter polar caps demonstrate similar response to solar wind dynamic pressure pulses, but the summer PC value is persistently higher than the winter PC value (up to factor 1.5). The predominant growth of the summer PC index is indicative of important role of ionospheric conductance in mechanism of the pressure pulses effect that implies the better conditions for field-aligned currents closure through the well-conducting sunlit ionosphere.

4. Relation of the PC index to magnetospheric substorms

Dynamics of magnetic disturbances in the auroral zone is described by the "auroral indices" AU and AL that characterize intensity of magnetic disturbances produced, respectively, by eastward and westward electric currents (electrojets) flowing in the morning and evening sectors of auroral zone. Their total, AE index, is regarded as a measure of disturbance in the auroral zone. During substorm periods a powerful westward electrojet is developed in the midnight auroral zone as a result of substorm current wedge formation (Birkeland, 1908) in response to strongly enhanced auroral particle precipitation and short-circuiting of neutral sheet currents through the high conductivity auroral ionosphere. That is why a sudden increase of the AL index is identified with the magnetospheric substorm onset, the intensity of substorm being evaluated by the AE or AL indices. The substorm sudden onset is usually preceded by gradual increase of westward and eastward electrojets regarded as a substorm growth phase. Growth phase is related to Region 1 FAC enhancement, which is accompanied by progressive intensification of auroral particle precipitation and formation Region 2 FAC system.

Relationship between the PC index and development of the isolated substorms, occurring against the background of magnetic quiescence, was analyzed by Janzhura et al. (2007). The following four classes of isolated magnetic disturbances were examined: weak magnetic bays, short magnetic substorms with duration under 3 hours, long substorms lasting more than 3 hours, and extended substorms, which demonstrate, after sudden onset a slow intensity increase, with maximum being retarded for some hours after the sudden onset.

Relationship between the PC index and progress of the strongest "sawtooth" substorms, whose intensity periodically increases and decreases, was analyzed by Troshichev & Janzhura (2009). The sawtooth events, developing under influence of a high-speed solar wind with strong fluctuating or a steady southward IMF, are distinguished from usual substorms by a larger local time extent (Lui et al., 2004; Henderson et al., 2006a,b; Clauer et al., 2006). Results (Troshichev et al., 2011b) indicate that a permanently high level of auroral activity is a typical feature of powerful sawtooth substorms. Aurora activity starts long before a magnetic disturbance onset and keeps a high level irrespective of magnetic disturbances. As a result, the close agreement between magnetic disturbance sudden onsets and behavior of aurora and particle injections on the synchronous orbit breaks down in contrast to "classical" substorms which are strongly associated with auroral particle precipitation.

The epoch superposition method, with the time of a substorm sudden onset taken as a key date, was used to reveal regularity in relationship between PC and AE indices in course of different types of substorm. In case of sawtooth substorms observed in 1998-2001 under conditions of the fluctuating southward IMF (N=43), the relationships between the IMF Bz and By components, electric field E_{KL}, and PC and AL indices are also examined. **Figure 6** integrates results obtained in analyses of Janzhura et al. (2007) and Troshichev & Janzhura (2009) for magnetic bays (average $AL < 200$ nT), short ($AL > 200$ nT), extended ($AL > 200$ nT) and sawtooth ($AL > 500$ nT) substorms (notice that the scale at the bottom panel is about

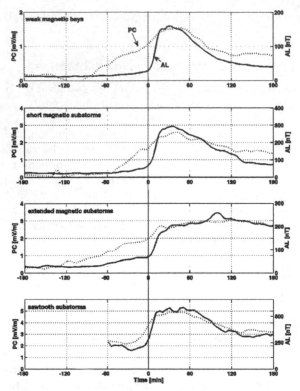

Fig. 6. Relationship between average PC and AL indices for weak magnetic bays (first panel), isolated short and extended substorms (second and third panels), and for sawtooth substorms (forth panel) (Troshichev and Janzhura, 2009).

twice as large as the scale for isolated substorms). One can see that the substorm sudden onset is preceded by the persistent PC index increase irrespective of substorm intensity. Following the generally accepted terminology, we shall name a time interval between a PC growth beginning and an explosive escalation of a magnetic disturbance in the auroral zone as a substorm growth phase. The mean duration of the growth phase is about 1 hour for isolated substorms and about 20-30 minutes for sawtooth substorms. As shown in (Troshichev and Janzhura, 2009) the PC growth phase is determined by the related E_{KL} function growth. In case of sawtooth substorms both E_{KL} and PC quantities stay at a high level after the substorm onset as well. As for magnetic disturbances in the auroral zone, the mean AL index keeps at low level before substorm and starts to grow only ~ 10 minutes ahead of the AL sudden onset.

These regularities clearly demonstrate that PC index variations in course of substorm are controlled by interplanetary electric field E_{KL} and that magnetic disturbances in the auroral zone follow the E_{KL} and PC increase. There is a definite PC level required for the substorm onset: in case of isolated substorms the threshold is about 1 – 2 mV/m, in case of sawtooth substorms the threshold is above 2 mV/m for. The fall of the PC value below 1 mV/m is unconditionally followed by the substorm decay.

Sawtooth disturbances last about 1.5-2 hours like isolated magnetic bays and short substorms, but they principally differ by PC behavior after the expansion phase. In case of isolated substorms, the PC index returns to a quiet level (<2 mV/m); in case of periodically repetitive substorms PC remains on a high level (>3 mV/m). If we consider the PC index as a signature of the solar wind energy input into the magnetosphere, an evident conclusion follows that isolated magnetic bays and short substorms are caused by a one-step energy income into the magnetosphere lasting for a short time, whereas isolated long and extended substorms take place if the solar energy is delivered with a different capability for a longer time. Sawtooth substorms are generated when a very powerful energy supply proceeds during a long period.

Thus, the analyses (Janzhura et al., 2007 and Troshichev & Janzhura, 2009) revealed that the PC index growth is a precursor of substorms development irrespective of a substorm type (isolated or sawtooth) and intensity. The growth phase duration is determined by the PC growth rate: the higher is the rate, the shorter is the growth phase duration. As **Figure 7** shows, the PC growth rate is a controlling factor for such important characteristics of magnetic disturbance as an AL growth rate before the substorm sudden onset (AL_{GR} = - 0.5 + 55*PC_{GR}, R=0.996) and for maximal intensity of magnetic substorm ($ALmax$ = 114 + 6570*PC_{GR}, R=0.997). According to last relationship, the $ALmax$ value will reach ~1600 nT when PC_{GR} is ~ 0.2 mV/m/min, which is equivalent to the PC jump by 2 mV/m per 10 minutes, that is observed in the case. Therefore, the average substorm intensity (i.e. magnitude of magnetic disturbance in the auroral zone) is predetermined by the PC growth rate during the growth phase.

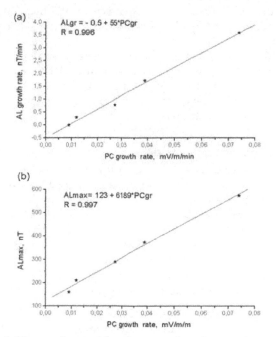

Fig. 7. Dependence of AL growth rate (a) and magnetic substorm intensity $ALmax$ (b) on $PCgrowth$ rate preceding a sudden onset (Troshichev and Janzhura, 2009).

The above presented results suggest that the PC index is a signature of solar wind energy that entered into the magnetosphere in course of solar wind–magnetosphere coupling. Indeed, the magnetospheric substorms are always preceded by the PC index growth and start as soon as the PC index exceeds a definite threshold value. Threshold value appears to be a level when the excess of energy incoming into the magnetosphere is over the ordinary energy dissipating in the magnetosphere; if threshold is not reached, the substorm can not begin, irrespective of how long the solar wind energy was income into the magnetosphere. From this point of view, the substorm growth phase is a period of the enhanced rate of energy pumping into the magnetosphere, not a period of energy storage in the magnetosphere. The large PC growth rate indicates that rate of energy pumping into the magnetosphere increases respectively. The greater the energy input, the faster is reached the threshold level; the shorter is the growth phase duration. The higher the energy input rate (i.e. energy input related to the growth phase duration), the larger is excess of incoming energy over the level of ordinary energy dissipation, and the more powerful is substorm. The succeeding substorm development is determined by dynamics of the subsequent input of energy, which is displayed by the PC index run. Thus, the solar wind geoefficiency in case of magnetospheric substorms can be monitored with confidence by the PC index.

5. Relation of the PC index to magnetic storms

Magnetic storms are the result of a joint action of magnetopause currents (DCF), which are proportional to the square root of the solar wind dynamic pressure and ring currents (DR) flowing in the inner magnetosphere (Chapman, 1963). The DR current ground effect typically far exceeds the DCF current effect, that is why the magnetic storm intensity is evaluated by the Dst index depicting a longitudinally averaged magnetic field depression at low latitudes (Sugiura, 1976). It is well known (Kamide, 1974; Russel et al., 1974; Burton et al., 1975; Akasofu, 1981) that magnetic storms intensity is dominantly controlled by southward IMF component (B_{ZS}), whereas the solar wind velocity (v) and density (n) are of minor importance. While investigating solar wind–magnetosphere coupling functions, the best result was obtained for functions including the geoeffective interplanetary electric field E_{KL} (Newell et al., 2008; Spencer et al., 2009).

Relationship between the 1-min PC index behavior and the storm depression development (Dst index) for epoch of solar maximum (1998–2004) has been examined by Troshichev et al. (2011c). Two criteria were used as a basic guideline to choose magnetic storms for the analysis: (1) magnetic storm duration should be longer than 12 hours, (2) magnetic storm depression should be larger than $Dst = -30$nT. On the basis of these criteria, 54 magnetic storms were separated for the period of 1998–2004 with a maximal storm intensity varying in the range from $Dst = -30$nT to $Dst = -373$nT. For convenience of comparison of the $PC(E_{KL})$ behavior with storm development, the main phase of each magnetic depression was divided into two parts: a "growth phase" when magnetic depression increases, and a "damping phase" when magnetic depression decreases, a "recovery phase" being used as before: as a period of a magnetic field slow restoration to the previous undisturbed level.

Since the Dst value is determined by a joint action of two, DCF and DR, current sources, the Dst index initial decline can be caused by a DR current growth as well as by a solar wind pressure reduction. In addition, during the solar maximum epoch (1998-2004), magnetic storms were usually following one after other, when a new magnetic storm started against the background of the recovery phase of the previous storm. Under these conditions, it would be well to look for another characteristic of storm depression beginnings, independent on the peculiarities of individual magnetic storm development. Referring to results (Janzhura *et al.*, 2007; Troshichev & Janzhura, 2009) and suggesting that a comparable input of the solar wind energy is required for development of magnetospheric substorms and magnetic storms, the value $PC=2mV/m$ was considered as a possible threshold level.

Examination of 54 chosen storms showed that all of them occurred under condition $PC>2$ mV/m. To demonstrate that a threshold level $PC=2$ mV/m is a typical feature of relationships between PC (or E_{KL}) changes and magnetic storm development, the magnetic storms were separated into six gradations according to their intensity, and the averaged $PC(E_{KL})$ and Dst quantities for these gradations were examined. The gradations determined by a minimal Dst value are the following: (a) $-30>Dst>-50nT$, (b) $-50>Dst>-80nT$, (c) $-80>Dst>-100nT$, (d) $-100>Dst>-120nT$, (e) $-130>Dst>-160nT$, and (f) $-160>Dst>-240nT$. The method of superposed epochs was used. Correspondingly, the time when the PC index

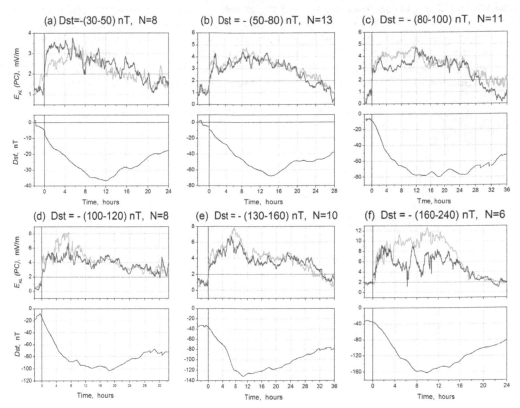

Fig. 8. Relationship between behavior of averaged PC index (red) and E_{KL} (green) quantities and development of magnetic storm Dst index (black) for 6 gradations of storm intensity: (a) $-30 > Dst > -50$ nT, (b) $-50 > Dst > -80$ nT, (c) $-80 > Dst > -100$ nT, (d) $-100 > Dst > 120$ nT, (e) $130 > Dst > 160$ nT, (f) $160 > Dst > 240$ nT. The key date (T=0) is taken as a time of the persistent transition of the E_{KL} value over the level of 2 mV/m (Troshichev et al., 2011c).

persistently rises above the level of 2 mV/m was taken as a time (T=0) of the disturbance beginning, and the time when the *PC* index persistently falls below the level of 2 mV/m was taken as a time of the recovery phase beginning.

Figure 8 shows the behavior of coupling function E_{KL} and the *PC* index, as well as corresponding changes in the *Dst* index for these 6 gradations. Results of case studies and statistical analysis (Figure 8) demonstrate the following regularities: geomagnetic field depression generally starts to develop as soon as *PC* and E_{KL} exceed the threshold of ~2 mV/m; as a rule, *PC* and E_{KL} simultaneously cross the threshold, although sometimes one goes ahead of other; when *PC* (E_{KL}) demonstrates repetitive strong enhancements and decreases, the magnetic storm displays the appropriate multiple depressions with growth and dumping phases; persistent descent of *PC* (E_{KL}) below the threshold level of 2 mV/m is indicative of the end of the storm main phase and transition to the recovery phase; "*PC* saturation effect" is typical of events with E_{KL} values > 6 mV/m.

To derive statistical relationships between *Dst* and mean *PC* (E_{KL}) values, the 1 min quantities *PC* and E_{KL} were averaged over the growth phase duration (the interval from time T=0 to the time of the peak value of *Dst* (*Dst(peak)*). The averaged values *PCgrowth* and $E_{KL}growth$ were compared with value of maximal depression. It turned out (Figure 9) that under conditions of E_{KL} < 6 mV/m, the relationship between *Dst(peak)* and *PC* (*Dst* = 24.8-31.8*PC) is of the same character as between *Dst(peak)* and E_{KL} (*Dst*=24.9-30.9*E_{KL}), although the correlation of *Dst* with E_{KL} (R=-0.74) is much lower than with *PC* (R=-0.87). Under

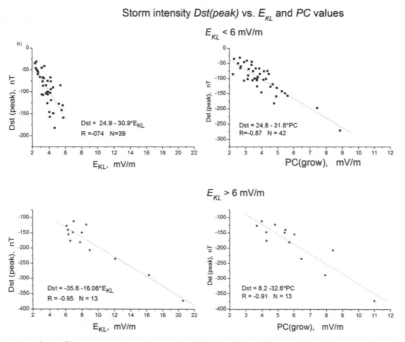

Fig. 9. Relationships between storm intensity *Dst(peak)* and quantities $E_{KL}growth$ and *PCgrowth* averaged over the storm growth phase interval, derived for categories of E_{KL} < 6 mV/m and E_{KL} > 6 mV/m (Troshichev et al., 2011c)

conditions $E_{KL}>6$ mV/m the efficiency of E_{KL} strongly decreases (Dst=-35.6-16.05*E_{KL}), whereas relationship between $Dst(peak)$ and PC is practically unchanged (Dst=8.2-32.6PC). The evident slowing down of depression $Dst(peak)$ for conditions of $E_{KL} > 5$-6 mV/m can be termined as a "Dst saturation effect" which is not seen in relations of Dst with the PC index.

Thus, all examined magnetic storms started as soon as the field E_{KL} and the appropriate PC index firmly exceeded the threshold of ~2 mV/m. The storm main phase lasted till the E_{KL} and PC steadily stands above this threshold level. The time of the firm descent of the PC and E_{KL} quantities below the same level of ~ 2 mV/m is indicative of the storm transition to the recovery phase. The storm intensity $Dst(peak)$ demonstrates the steady linear correlation with the PC value averaged for the growth phase $PC(growth)$ for all storm intensities irrespective of PC value. The storm dynamics correlate better with value and changes of the PC index than with those of the E_{KL} field that provides a weighty argument in support of the PC index as a signature of the solar wind energy that entered into the magnetosphere.

6. Conclusion

The PC index has been introduced initially (Troshichev et al., 1988) as a characteristic of the polar cap magnetic activity related to the geoeffective interplanetary electric field E_{KL} which was determined by formula of Kan & Lee (1979). The recent studies (Troshichev et al., 2007; Janzhura et al., 2007, Troshichev & Janzhura, 2009; Troshichev et al., 2011c) show that the PC index implication is more significant: the magnetospheric storms and substorms start only if the PC index reaches the definite threshold value (~ 2 mV/m for storms, and >1.5mV/m for substorms); the substorm growth phase duration and substorm intensity are determined by the PC growth rate and substorms are stopped as soon as PC index falls below 1-1.5 mV/m; the storm length is terminated by duration of period, when $PC>2$mV/m, the storm intensity being linearly related to the PC index averaged for the storm time interval; development of storms and substorms is better consistent with the PC behavior than with the coupling function variations; and so on. In addition, it turned out that the PC index adequately responds to sharp changes in the solar wind dynamic pressure. All these experimentally established relationships make it possible to conclude that the PC index should be regarded as an adequate and convenient proxy of the solar wind energy that entered into the magnetosphere.

If holding this point of view it becomes obvious why the storm and substorm indicators should correlate better with the PC index than with coupling function E_{KL} (since the coupling function E_{KL} characterizes the state of solar wind coupling with the magnetosphere, whereas the PC index characterizes the energy that entered into the magnetosphere); why the PC index increase precedes storm and substorms (because growth of the entered energy above the level of energy dissipation is followed by realizing the energy excess in form of magnetic disturbances); why the sawtooth substorms demonstrate the distinct periodicity under conditions of steadily high energy supply (because the persistently high entered energy ensures the extreme intensity of field aligned currents discharging the current generator which power is limited by finite plasma pressure gradients in the closed magnetosphere (see Troshichev et al., 2012)).

As far as the PC index characterizes the energy that entered into the magnetosphere in course of solar wind – magnetosphere coupling, the index can be used to monitor the solar wind geoefficiency and state of magnetosphere. Since disturbances in magnetosphere are always preceded by energy input, the PC index usage makes it possible to realize the space weather nowcasting (including the auroral ionosphere state and even the anomalous processes in polar atmosphere). At present, the PCN and PCS indices are published at sites:

PCN ftp://ftp.space.dtu.dk/WDC/indices/pcn/

PCS http://www.geophys.aari.ru/

7. References

Akasofu, S.-I. (1981). Energy coupling between the solar wind and the magnetosphere, *Space Sci. Rev.*, vol.28, 121

Antonova, E.; Kirpichev I. & Stepanova, M. (2006). Field-aligned current mapping and the problem of the generation of magnetospheric convection. *Adv. Space Res.*, vol.38, 1637 – 1641, doi:10.1016/j.asr.2005.09.042

Antonova, E,.; Kirpichev, I.; Ovchinnikov, I.; Pulinets, M.; Znatkova, S.; Orlova, K. & Stepanova, M. (2011). Topology of High-Latitude Magnetospheric Currents. In: Fujimoto M, Liu W (eds) *The Dynamic Magnetosphere*. IAGA Special Sopron Book Series, V 3, Springer Dordrecht Heidelberg London New York, pp.201-210

Arnoldy, R. (1971). Signature in the interplanetary medium for substorms. *J. Geophys. Res.*, vol.76, 5189-5201

Baliff, J.; Jones, D.; Coleman, P.; Davis, L. & Smith, E. (1967). Transverse fluctuations in the interplanetary magnetic field: a requisite for geomagnetic variability. *J. Geophys. Res.*, vol.72, 4357-4362

Birkeland, K. (1908). *The Norwegian Aurora Polaris Expedition 1902-1903,* vol.1 Christiania

Burton, R.; McPherron, R. & Russell, C. (1975). An empirical relationship between interplanetary conditions and Dst. *J. Geophys. Res.*, vol.80, 4204-4214

Bythrow, P. & Potemra, T. (1983). The relationship of total Birkeland currents to the merging electric field. *Geophys Res. Lett.*, vol.10, 573-576

Chapman, S. (1963). Solar plasma, geomagnetism and aurora, *in Geophysics: The Earth's Environment*. C.DeWitt, (Ed), Gordon and Breach Sci.Pub., New York - London

Clauer, C.; Cai, X.; Welling, D.; DeJong, A. & Henderson, M. (2006). Characterizing the 18 April 2002 storm-time sawtooth events using ground magnetic data. *J. Geophys. Res.*, vol.111, A04S90, doi:10.1029/2005JA011099

Fairfield, D. & Cahill, L. (1966). Transition region magnetic field and polar magnetic disturbances. *J. Geophys. Res.*, vol.71, 6829-6846

Foster, J.; Fairfield, D.; Ogilvie, K. & Rosenberg, T. (1971). Relationship of interplanetary parameters and occurrence of magnetospheric substorms. *J. Geophys. Res.*, vol.76, 6971-6975

Garrett, H.; Dessler, A. & Hill, T. (1974). Influence of solar wind variability on geomagnetic activity. *J. Geophys. Res.*, vol.79, 4603-4610

Gizler, V.; Semenov, V. & Troshichev, O. (1979). The electric fields and currents in the ionosphere generated by field-aligned currents observed by TRIAD. *Planet. Space Sci.*, vol.27, 223-231

Henderson, M.; Reeves, G.; Skoug, R.; Thomsen, M.; Denton, M.; Mende, S.; Immel, T.; Brandt, P. & Singer, H. (2006a) Magnetospheric and auroral activity during the 18 April 2002 sawtooth event. *J. Geophys. Res.*, vol.111, A01S90, doi:10.1029/2005JA011111

Henderson, M.; Skoug, R.; Donovan, E; et al. (2006b). Substorm during the 10 August 2000 sawtooth event. *J. Geophys. Res.*, vol.111, A06206, doi:10.1029/2005JA011366

Hirshberg, J. & Colburn D. (1969). Interplanetary field and geomagnetic variations – a unified view. *Planet. Space Sci.*, vol.17, 1183-1206

Iijima, T. & Potemra, T. (1976). The amplitude distribution of field-aligned currents at northern high latitudes observed by Triad. *J. Geophys. Res.*, vol.81, 2165-2174

Iijima, T. & Potemra, T. (1978). Large-scale characteristics of field-aligned currents associated with substorms. *J. Geophys. Res.*, vol.83, 599-615

Iijima, T. & Potemra, T. (1982). The relationship between interplanetary quantities and Birkeland current densities. *Geophys. Res. Lett.*, vol.4, 442-445

Iijima, T.; Potemra, T. & Zanetti, L. (1997). Contribution of pressure gradients to the generation of dawnside region 1 and region 2 currents. *J. Geophys. Res.*, vol.102: 27069-27081

Janzhura, A. & Troshichev, O. (2008). Determination of the running quiet daily geomagnetic variation, *J. Atmos. Solar-Terr. Phys.*, vol.70, 962–972

Janzhura, A.; Troshichev, O. & Stauning, P. (2007). Unified PC indices: Relation to the isolated magnetic substorms, *J. Geophys. Res.*, vol.112, A09207, doi: 10.1029/2006JA012132

Kamide, Y. (1974). Association of DP and DR fields with the interplanetary magnetic field variations. *J. Geophys. Res.*, vol.79, 49

Kan, J. & Lee, L. (1979). Energy coupling function and solar wind-magnetosphere dynamo. *Geophys. Res. Lett.*, vol.6, 577-580

Kane, R. (1974). Relationship between interplanetary plasma parameters and geomagnetic Dst. *J. Geophys. Res.*, vol. 79, 64-72

Kokubun, S. (1972). Relationship of interplanetary magnetic field structure with development of substorm and storm main phase. *Planet. Space Sci.*, vol.20, 1033-1050

Kuznetsov, B. & Troshichev O. (1977). On the nature of polar cap magnetic activity during undisturbed periods. *Planet. Space Sci.*, vol.25, 15-21

Langel, R. (1975). Relation of variations in total magnetic field at high latitude with parameters of the IMF and with DP2 fluctuations. *J. Geophys. Res.*, vol.80, 1261-1270

Lui, A.; Hori, T.; Ohtani, S.; Zhang, Y.; Zhou, X. Henderson, M.; Mukai, T.; Hayakawa, H. & Mende, S. (2004). Magnetotail behavior during storm time 'sawtooth injections'. *J. Geophys. Res.*, vol.109, A10215, doi:10.1029/2004JA010543

Maezawa, K. (1976). Magnetospheric convection induced by the positive and negative Z components of the interplanetary magnetic field: quantitative analysis using polar cap magnetic records. *J. Geophys. Res.*, vol.81, 2289-2303

Mansurov, S. (1969). A new evidence for relationship between the space and earth magnetic fields. *Geomagn. Aeronomy*, vol.9, 768-770 (in Russian)

McDiarmid, I.; Budzinski, E.; Wilson, M. & Burrows J. (1977). Reverse polarity field-aligned currents at high latitudes. *J. Geophys. Res.*, vol.82, 1513-1518

Meng, C-I.; Tsurutani, B.; Kawasaki, K. & Akasofu, S-I. (1973). Cross-correlation analysis of the AE-index and the interplanetary magnetic field Bz component. *J. Geophys. Res.*, vol.78, 617-629

Murayama, T. & Hakamada, K. (1975). Effects of solar wind parameters on the development of magnetospheric substorms. *Planet. Space Sci.*, vol.23, 75-91

Newell, P.; Sotirelis, T.; Liou, K.; Meng, C-I. & Rich, F. (2007) A nearly universal solar wind magnetosphere coupling function inferred from 10 magnetospheric state variables. *J. Geophys. Res.*, vol.112, A01206, doi:10.1029/2006JA012015

Newell, P.; Sotirelis, T.; Liou, K. & Rich, F. (2008). Pairs of solar wind-magnetosphere coupling functions: combining a merging term with a viscous term works best. *J. Geophys. Res.*, vol.113, A04218, doi:10.1029/2007JA012825

Nisbet, J.; Miller, M. & Carpenter, L. (1978). Currents and electric fields in the ionosphere due to field-aligned auroral currents. *J. Geophys. Res.*, vol.83, 2647

Nishida, A. (1968). Geomagnetic DP2 fluctuations and associated magnetospheric phenomena. *J. Geophys. Res.*, vol.73, 1795-1803

Perreault, P. & Akasofu, S-I. (1978). A study of geomagnetic storms. *Geophys. J. R. Astr. Soc.*, vol.54, 57

Pudovkin, M.; Raspopov, O.; Dmitrieva, L.; Troitskaya, V. & Shepetnov, R. (1970). The interrelation between parameters of the solar wind and the state of the geomagnetic field. *Ann. Geophys.*, vol.26,: 389-392

Reiff, P. & Luhmann, J. (1986). Solar wind control of the polar-cap voltage. In: *Solar Wind – Magnetosphere Coupling*, Kamide, Y. & Slavin, J. (eds), pp 453-476, Terra Sci, Tokyo

Rossolenko, S.; Antonova, E.; Yermolaev, Y.; Verigin, M.; Kirpichev, I. & Borodkova, N. (2008). Turbulent Fluctuations of Plasma and Magnetic Field Parameters in the Magnetosheath and the Low-Latitude Boundary Layer Formation: Multisatellite Observations on March 2, 1996. *Cosmic. Res.*, vol. 46, 373–382 (in Russian).

Rostoker, G. & Falthammar, C. (1967). Relationship between changes in the interplanetary magnetic field and variations in the magnetic field at the Earth's surface. *J. Geophys. Res.*, vol.72, 5853-5863

Russel, C.; McPherron, R. & Burton, R. (1974). On the cause of geomagnetic storms. *J. Geophys. Res.*, vol.79, 1105

Sergeev, V. & Kuznetsov, B. (1981) Quantitative dependence of the polar cap electric field on the IMF BZ component and solar wind velocity. *Planet. Space Sci.*, vol.29, 205-213

Spencer, E.; Rao, A.; Horton, W. & Mays, M. (2009). Evaluation of solar – magnetosphere coupling functions during geomagnetic storms with the WINDMI model. *J. Geophys. Res.*, vol.114, A02206, doi: 10.1029/2008JA013530

Stepanova, M.; Antonova, E., Bosqued, J. & Kovrazhkin, R. (2004). Azimuthal plasma pressure reconstructed by using the Aureol-3 satellite date during quiet geomagnetic conditions. *Adv. Space Res.*, Vol. 33, 737-741

Stepanova, M.; Antonova, E. & Bosqued J. (2006). Study of plasma pressure distribution in the inner magnetosphere using low-altitude satellites and its importance for the large-scale magnetospheric dynamics, *Adv. Space Res.*, Vol. 38, 1631-1636

Sugiura, M. (1976). Hourly values of equtorial Dst for the IGY, *Ann. Int.Geophys.Year*, 35, 1.

Svalgaard, L. (1968). Sector structure of the interplanetary magnetic field and daily variation of the geomagnetic field at high latitudes. *Det Danske meteorologiske institute Charlottenlund*, preprint R-6

Troshichev, O. (1975). Magnetic disturbances in polar caps and parameters of solar wind. In: *Substorms and magnetospheric disturbances*. Nauka, Leningrad, pp 66-83, (in Russian)

Troshichev, O. (1982). Polar magnetic disturbances and field-aligned currents. *Space Sci. Rev.*, vol. 32, 275-360

Troshichev, O. & Tsyganenko, N. (1978). Correlation relationships between variations of IMF and magnetic disturbances in the polar cap. *Geomagn. Res.*, vol.25, 47-59 (in Russian)

Troshichev, O. & Andrezen, V. (1985). The relationship between interplanetary quantities and magnetic activity in the southern polar cap. *Planet. Space Sci.*, vol.33, 415

Troshichev, O. & Janzhura, A. (2009). Relationship between the PC and AL indices during repetitive bay-like magnetic disturbances in the auroral zone. *J. Atmos. Solar-Terr. Phys.*, 71, 1340–1352

Troshichev, O.; Gizler, V.; Ivanova, I. & Merkurieva, A. (1979). Role of field-aligned currents in generation of high latitude magnetic disturbances. *Planet. Space Sci.*, vol.27, 1451-1459

Troshichev, O.; Andrezen, V.; Vennerstrøm, S. & Friis-Christensen, E. (1988). Magnetic activity in the polar cap – A new index. *Planet. Space Sci.*, vol.36, 1095,.

Troshichev, O.; Janzhura, A. & Stauning, P. (2006). Unified PCN and PCS indices: Method of calculation, physical sense and dependence on the IMF azimuthal and northward components. *J. Geophys. Res.*, vol.111, A05208, doi:10.1029/2005JA011402

Troshichev, O.; Janzhura, A. & Stauning, P. (2007). Magnetic activity in the polar caps: Relation to sudden changes in the solar wind dynamc pressure, *J. Geophys. Res.*, vol.112, A11202, doi:10.1029/2007JA012369,

Troshichev, O.; Podorozhkina, N. & Janzhura, A. (2011a). Invariability of relationship between the polar cap magnetic activity and geoeffective interplanetary electric field. *Ann. Geophys.*, vol.29, 1–11, doi:10.5194/angeo-29-1-2011

Troshichev, O.; Stauning, P.; Liou, K. & Reeves, G. (2011b). Saw-tooth substorms: inconsistency of repetitive bay-like magnetic disturbances with behavior of aurora. *Adv. Space Res.*, vol.47, 702–709, doi: 10.1016/j.asr.2010.09.026

Troshichev, O.; Sormakov, D. & Janzhura, A. (2011c). Relation of PC index to the geomagnetic storm Dst variation. *J. Atmos. Solar-Terr. Phys.*, vol.73 611–622, doi:10.1016/j.jastp.2010.12.015

Troshichev, O.; Sormakov, D. & Janzhura, A. (2012). Sawtooth substorms generated under conditions of the steadily high solar wind energy input into the magnetosphere: Relationship between *PC*, *AL* and *ASYM* indices. *Adv. Space Res.*, doi:10.1016/j.asr.2011.12.011

Wilcox, J.; Schatten, K. & Ness, N. (1967). Influence of interplanetary magnetic field and plasma on geomagnetic activity during quiet-sun conditions. *J. Geophys. Res.*, vol.72, 19-26

Wing, S. & Newell, P. (2000). Quiet time plasma sheet ion pressure contribution to Birkeland currents. *J. Geophys. Res.*, vol.105, 7793-7802

Xing, X.; Lyons, R,; Angelopoulos, V.; Larson, D; McFadden, J.; Carlson, C.; Runov, A. & Auster, U. (2009). Azimuthal plasma pressure gradient in quiet time plasma sheet. *Geophys. Res. Lett.*, vol.36, L14105, doi:10.1029/2009GL038881

Zmuda, A. & Armstrong, J. (1974). The diurnal flow pattern of field-aligned currents. *J. Geophys. Res.*, vol.79, 4611-4519

8

Solar Wind Sails

Ikkoh Funaki[1] and Hiroshi Yamakawa[2]
[1]Japan Aerospace Exploration Agency
[2]Kyoto University
Japan

1. Introduction

Magnetic sail (MagSail) is a unique but never-realized interplanetary propulsion system. The original idea of MagSail proposed by Zubrin is depicted in Fig.1 (Andrews & Zubrin, 1990). A MagSail spacecraft has a hoop coil, and it produces an artificial magnetic field to reflect the solar wind particles approaching the coil, and the corresponding repulsive force exerts on the coil to accelerate the spacecraft in the solar wind direction. MagSail and its derivatives are usually called as solar wind sail in contrast to solar light sail, which is propelled by the solar light pressure. Before realizing the ideas of solar wind sail, however, fundamental studies are required from both physical and engineering points of view, in particular, experimental validation of these ideas are very important before applying these concepts to realistic spacecraft design.

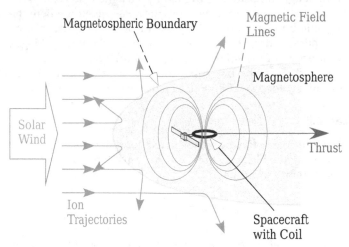

Fig. 1. Solar Wind Sail (MagSail).

To simulate a solar wind sail in laboratory, we fabricated a plasma wind tunnel facility (Funaki, 2007). Our special effort in laboratory experiment is directed to satisfy the similarity law associated with plasma flows of MagSail and its derivatives. Also, the laboratory experiment is intended for direct thrust measurement, which was never tried in the past. In this chapter, after introducing the basics of sail propulsion using the solar wind, details of experimental facility and experimental results of MagSail in laboratory are described.

2. Principle of MagSail

In this section, the idea and feature of MagSail are introduced and its thrust formula is provided.

2.1 Original MagSail by Zubrin

In 1991, Zubrin released the idea of interplanetary MagSail, which deploys superconducting cable after launched into space (Zubrin & Andrew, 1991). Before the launch of a MagSail spacecraft, a loop of superconducting cable, which is millimeter in diameter is attached on a drum onboard the spacecraft. After launch, the cable is released from the spacecraft to form a loop of tens of kilometers in diameter, then, a current is initiated in the loop. Once the current is initiated, it will be maintained in the superconductor without operating power supply. The magnetic field created by the current will impart a hoop stress to the loop for aiding the development and eventually forcing it to a rigid circular shape.

When MagSail is operated in interplanetary space, charged particles approaching the current loop are decelerated/deflected according to the B-field they experience. If the interacting scale length between the plasma flow and the magnetic field is large as in the cases of some magnetized planets like Earth and Jupiter, a magnetosphere (a magnetic cavity or a magnetic bubble) is formed around the current loop (Fig. 2). The solar wind plasma flow and the magnetic field are divided by a magnetopause, at which ions entering the magnetic field are reflected except near the polar cusp region where the ions can enter deep into the magnetic bubble. Due to the presence of the magnetosphere, the solar wind flow is blocked, creating a drag force exerting on the loop; thus a MagSail spacecraft is accelerated in the direction of the solar wind. The solar wind in the vicinity of the Earth is a flux of 10^6 protons and electrons per cubic meter at a velocity of 400 to 600 km/s. The maximum speed available for MagSail would be that of the solar wind itself.

2.2 Thrust of MagSail

One can easily imagine that a force on the current loop depends on the area blocking the solar wind. By increasing the blocking area, a large thrust force is expected. The force on the coil of a MagSail is therefore formulated as,

$$F = C_d \frac{1}{2} \rho_{sw} u_{sw}^2 S \tag{1}$$

where C_d is thrust coefficient, $1/2\rho_{sw}u_{sw}^2$ is the dynamic pressure of the solar wind, and S is the blocking area. In Eq.(1), $\rho_{sw}=m_i n$ is the density of the solar wind, m_i is the mass of an ion, n is the number density, and u_{sw} is the velocity of the solar wind.

To calculate F from Eq.(1), the blocking area S should be specified, but S is not priori given. It is therefore customary to employ the characteristic length, L, and to approximate the blocking area as $S=\pi L^2$. In this paper, we choose the standoff distance as the characteristic

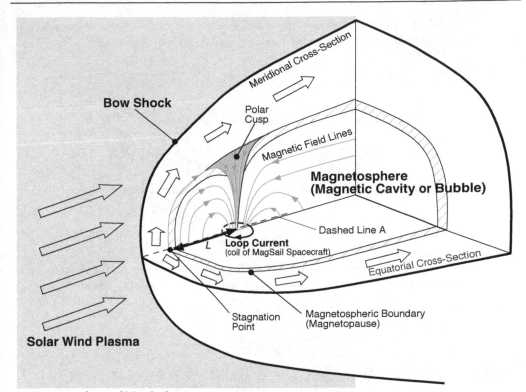

a) Magnetosphere of MagSail

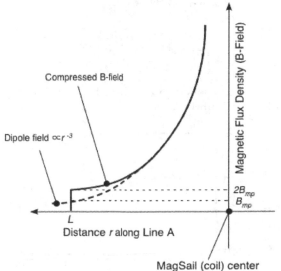

b) B-field distribution along stagnation line (dashed line A in a))

Fig. 2. Expected plasma and magnetic fields of MagSail.

length, which is the distance between the stagnation point and the center of the coil. L is derived from a pressure balance at the stagnation point (Obayashi, 1970) as

$$n\,m_i\,u_{sw}{}^2 = \frac{\left(2B_{mp}\right)^2}{2\mu_0}, \tag{2}$$

where B_{mp} is expressed using the magnetic moment, M_m, as

$$B_{mp} = \frac{\mu_0 M_m}{4\pi L^3}. \tag{3}$$

From these two equations, the standoff distance, L, is directly obtained as,

$$L = \left(\frac{\mu_0 M_m{}^2}{8\pi^2\,n\,m_i\,u_{sw}{}^2}\right)^{1/6}. \tag{4}$$

Because L is uniquely determined from solar wind properties as well as the magnetic moment of MagSail, L is a reasonable choice for the characteristic length of the flow around MagSail.

Using eq.(1) and $S=\pi L^2$, the correlation between L and F is derived as

$$F = C_d \frac{1}{2}\rho u_{sw}^2 \pi L^2 \tag{5}$$

in which C_d in Eq.(5) is a fitted curve to numerical simulation results (Fujita, 2004):

$$C_d = \begin{cases} 3.6\exp\left(-0.28\left(\dfrac{r_{Li}}{L}\right)^2\right) & \text{for} \quad \dfrac{r_{Li}}{L} < 1 \\[4mm] \dfrac{3.4}{\left(r_{Li}/L\right)}\exp\left(-0.22\left(\dfrac{L}{r_{Li}}\right)^2\right) & \text{for} \quad \dfrac{r_{Li}}{L} \geq 1 \end{cases} \tag{6}$$

where r_{Li} is ion's Larmor radius defined afterwards. Thrust by Eqs.(5) and (6) is plotted in Fig.3 The MagSail by Zubrin (Zubrin & Andrews, 1991) required a spacecraft with a large

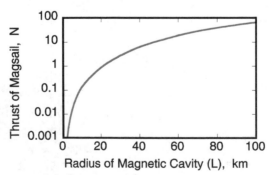

Fig. 3. Theoretical prediction of thrust by MagSail.

hoop coil of 64 km in radius to form 100-km-radius blocking area (which corresponds to approximately 30-N-class thrust). However, the dimension of the hoop coil was too large to realize. We therefore try to design a smaller MagSail than Zubrin proposed. From Fig. 3, one may notice that for L=20 km, 1 N is obtained. We selected the target of MagSail as 1 N, which corresponds to L=20 km; this is an adequate thrust level for medium-size (1,000-kg-class) deep space explorers.

2.3 Scaling parameters of MagSails

Empirical equation (6) indicates that thrust by MagSail is significantly influenced by the non-dimensional parameter, r_{Li}/L. In this section, similarity law for MagSail is explained after introducing two important scaling lengths, ion Lamor radius and skin depth.

2.3.1 Ion Larmor radius and skin depth

The boundary between a solar wind plasma and a magnetosphere is called magnetospheric boundary or magnetopause, where induced currents divide the plasma and the magnetic field regions. Magnetospheric boundary is schematically plotted in Fig.4. As in Fig.4, when charged particles impinge on the magnetospheric boundary from outside the magnetospheric boundary, ions can enter deep into the magnetosphere, whereas electrons are trapped and they are reflected back to the sun direction at the surface of the boundary. It is expected that the penetration depth of ions is comparable to ion Larmor radius:

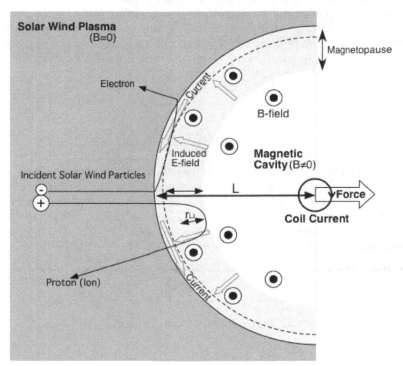

Fig. 4. Charged particle behaviour near magnetopause.

$$r_{Li} = \frac{m_i u_{sw}}{e \cdot 2B_{mp}} \qquad (7)$$

from which r_{Li}=72 km if we put m_i=1.67x10^{-27} kg, u_{sw}=400 km/s, e=1.6x10^{-19} C, and B_{mp}=29 nT.

Disparity between ion's and electron's penetration depth at this boundary will also cause charge separation; this sheath region will induce an outward electric field. By this electric field, an electron is accelerated to a high velocity, resulting in a motion with large Larmor radius in the magnetosphere. Such electron motion dominates magnetospheric current at the outer edge of the magnetopause. In contrast to the electron motion at the boundary, ions are decelerated by the same electric field. The theoretical length of charge separation is the skin depth:

$$\delta = c/\omega_p \qquad (8)$$

which is the decaying length of an electromagnetic wave incident on a plasma (Bachynski & Osborne, 1965); in eq.(8), c is the light speed and ω_p is the plasma frequency. For the solar wind plasma, $\delta \sim 1$ km. Realistic penetration depth takes a value between δ and r_{Li} when background neutralizing electrons reduce the magnitude of the outward electric field (Nishida, 1982).

2.3.2 Non-dimensional scaling parameters of MagSail

In the following, four non-dimensional parameters and similarity law for MagSail in space are obtained in analogy to the similarity law of geo-magnetophysics (Bachynski & Osborne, 1965). The solar wind is a super sonic plasma flow, and it consists of collisionless particles: typical mean free path of the solar wind is about 1 AU. These features are described by high Mach number, $M > 1$ as well as very high magnetic Reynolds number, $Rm \gg 1$. M and Rm are defined as follows:

$$M = \frac{u_{sw}}{\sqrt{\gamma R T_{sw}}} \qquad (9)$$

$$Rm = \sigma \mu_0 u_{sw} L \qquad (10)$$

where σ is electric conductivity, R is gas constant, γ is specific heat ratio, and μ_0 is permeability in vacuum. Putting typical plasma velocity and temperature of the solar wind (u_{sw}=400 km/s and $T_i \sim 10$ eV) into the above equation, $M \sim 8$ is obtained. Also, the assumption of Coulomb collision gives σ=2x10^4/Ωm and $Rm \sim 10^8$ for L=10 km. In addition to these two non-dimensional parameters, we defined r_{Li}/L, and δ/L, hence four non-dimensional parameters in total are introduced and they are listed in Table.1. Among them,

Parameters	MagSail	
	in space	in laboratory
Mach number	~ 8	> 1
Ratio of ion larmor radius to L (r_{Li}/L)	~ 1	~ 1
Ratio of skin depth to L (δ/L)	< 1	< 1
Magnetic Reynolds' number (Rm)	~10^8	~ 10

Table 1. Non-dimensional parameters of MagSail.

the parameters Rm, r_{Li}/L, and δ/L, are functions of the size of the magnetosphere, in which L was selected as 10 km<L< 100 km in our study. Corresponding non-dimensional parameters are 0.72 <r_{Li}/L < 7.2 (the ion gyration radius is comparable to or larger than L), which is in contrast to the MHD scale requiring r_{Li}/L<<1), and lastly, δ/L < 0.03 (the skin depth is much smaller than L).

Since δ <<r_{Li}, the effective thickness of magnetopause is considered to be r_{Li}. As shown in Fig.5, if the thickness of the magnetopause is small enough in comparison to L, almost all of the incident ions are reflected at the magnetospheric boundary, hence large thrust on the coil of the MagSail is expected. Vice versa, if the thickness of the magnetopause is much larger than L, no interaction between the plasma flow and the magnetic field is anticipated. We treat a transitional region from the MHD scale (thin magnetopause mode) to the ion kinetic scale (thick magnetopause mode) in this experiment.

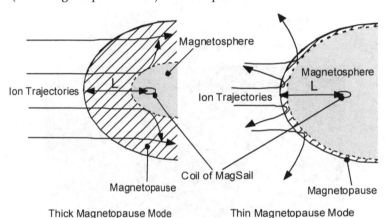

Fig. 5. Thick and thin magnetopause of MagSail.

3. Experimental setup for MagSail

Based on the above scaling consideration, experimental setup for MagSail is developed (Funaki, et al., 2006). The setup consists of a high-power solar wind simulator and a coil simulating MagSail spacecraft, both of which are operated in a quasi-steady mode of about 0.8-ms duration. In our experiment of MagSail (Fig.6), a coil simulating MagSail spacecraft (20-75 mm in diameter) was located at a downstream position of the solar wind simulator, and a plasma jet was introduced to a magnetic field produced by the coil. Typical snapshot is shown in Fig.7, in which plume jet as well as the coil simulating MagSail can be found.

A magnetoplasmadynamic (MPD) arcjet in Fig.8 was used as the solar wind simulator (SWS). The discharge chamber of the MPD arcjet is 50 mm in inner diameter and 100 mm in length. The MPD arcjet consists of eight molybdenum anode rods (8 mm in diameter) azimuthally located, and a short thoriated tungsten cathode rod 20 mm in diameter; they are surrounded by an annular floating body and insulators. These electrodes enable stable operate from a low current discharge range to an erosive high current discharge range. The SWS is attached on the space chamber inner wall as shown in Fig. 6.

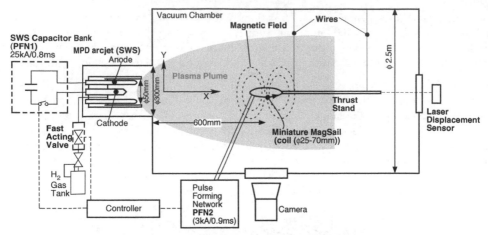

Fig. 6. Schematics of MagSail experimental facility.

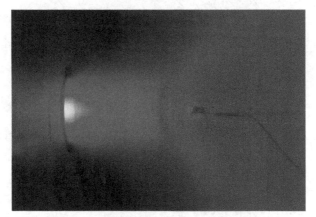

Fig. 7. Experiment of MagSail; typical snap shot.

Fig. 8. Magnetoplasmadynamic(MPD) arcjet as solar wind simulator (SWS).

A fast-acting valve (FAV) allowed us to feed gaseous propellants to the MPD arcjet. The hydrogen mass flow rate was controlled by adjusting the reservoir pressures. Timing of experiment is explained in Fig.9. After a gas pulse reaches its quasi-steady state, a pulse-forming network (PFN) is triggered. The PFN for SWS supplies the discharge current up to 20 kA with a flat-topped waveform in quasi-steady mode as shown in Fig.9. PFN for the coil (PFN2) supplies rather small current below 3 kA, and 20- to 75-mm-diameter 20-turn coil is required to produce up 2.0-T B-field at the center of the coil. After coil current (MagSail operation) reaches its steady state, discharge of the SWS (MPD arcjet) is initiated.

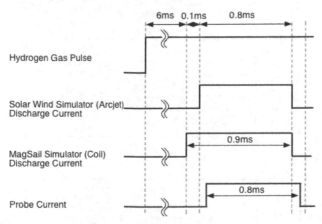

Fig. 9. Timing of MagSail experiment.

4. Experimental results of MagSail

In this section, after operational characteristics of MagSail in laboratory are described, thrust measurement results are provided. Then, the similarity law on a plasma flow of MagSail is discussed.

4.1 Operation of MagSail in lab

Discharge current profiles of the solar wind simulator (SWS) MPD arcjet and the coil are plotted in Fig.10. Flat-topped quasi-steady discharge continues about 0.8 ms in the case of SWS, and 1.0 ms in the case of the coil simulating MagSail spacecraft. High speed photos of MagSail experiment are shown in Fig.11. After the coil current (MagSail operation) reach its steady state, discharge of the MPD arcjet is initiated (Fig.11(a)). In Figs.11(b) and (c), it is observed that a wave front expands radially. This shock wave is followed by a quasi-steady plasma flow, that interacts with the magnetic field created by the solenoidal coil (Fig.11(d)).

During the quasi-steady interaction (0.25-1.0 ms in Fig.10), it was found that the plasma flow at the coil position was fluctuating. Averaged plasma parameters during the quasi-steady interaction were evaluated by Langmuir probe diagnostics as well as time-of-flight velocity measurement at the coil position. Results are listed in Table.2. The size of SWS's plume is 0.7 m in diameter (FWHM) at the coil position, and it is adequate for the laboratory experiment of Magnetic sail, which has typically 10-cm-size magnetic cavity. Scaling parameters can be calculated from the measurement in Table.2, but they are discussed afterwards in this chapter.

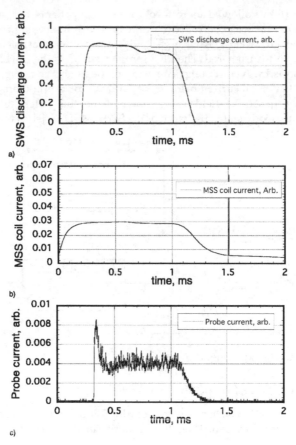

Fig. 10. Operation of solar wind simulator; a) discharge current profile of SWS, b) coil current profile, and c) plasma plume probe current profile (ion saturation current) at the coil position for H₂ 0.4 g/s, charging voltage of PFN for SWS is 4 kV, and charging voltage of PFN for coil is 1.5 kV.

Plasma stream form hydrogen MPD solar wind simulator	
Velocity	20–45 km/s
Plasma density	10^{18}–10^{19} /m³
Electron temperature	1 eV
Radius of plasma stream at the coil position	0.2–0.35 m
Plasma duration	0.8 ms
Coil current simulating MagSail in operation	
Radius of coil	9-37.5 mm
B-field at the center of coil	0–2.0 T
Duration of exciting current	0.9 ms

Table 2. Operating conditions and plasma parameters of solar wind simulator and coil simulating MagSail spacecraft.

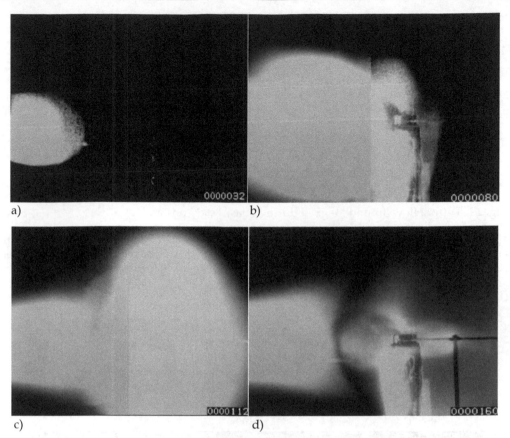

Fig. 11. High speed photos of pure MagSail experiment; a) just after initiating SWS (t=32 μs has passed after the discharge is initiated), b) t=80 μs, c) t=112 μs, and d) t=160 μs; d) corresponds to a quasi-steady state operation) in the case of 45 km/s, 2x10^{19} m^{-3} plasma flow from SWS, B-field at the center of coil is 1.8 T.

Another view of MagSail experiment is shown in Fig.12, in which a shutter camera was used to capture MagSail and its flowfield during a quasi-steady interaction. The most important feature of the interaction is the magnetospheric boundary between the SWS plasma flow and the low-energy plasma in the magnetic cavity. At the location of magnetospheric boundary, significant change of the magnetic field strength was found by a magnetic field measurement using a magnetic probe (Ueno, et al, 2009).

As far as we see Fig.12, the interaction seems very stable during quasi-steady operation of MagSail. However, if we see them with a high-speed camera, oscillatory magnetic cavity was observed (Oshio, et al., 2007; Oshio, et al., 2011). Thrust is hence produced under turbulent environment. Figure 13 shows high-speed photos when the discharge current of the MPD arcjet (J_{SWS}) is 11.6 kA. In Fig. 13, t = 0 corresponds to the time initiating the discharge of the SWS, and the time difference between each frame photograph is 2 μs, and shutter time is 0.5 μs. The interaction has already reached quasi-steady state in Fig. 13, in

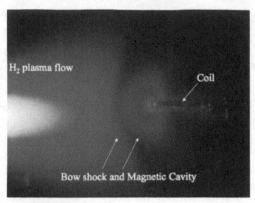

Fig. 12. Typical flow around MagSail during quasi-steady operation.

spite, an oscillating magnetosphere and plasma plume was observed. For example, when comparing the photos at 506 μs and at 516 μs in Fig. 13, the magnetospheric boundary is different. Also, when we see them in a longer time scale, it is found that the magnetospheric size shrinks and expands by about 20% repeatedly, and the averaged magnetospheric size is 0.15 m in the case of Fig. 13.

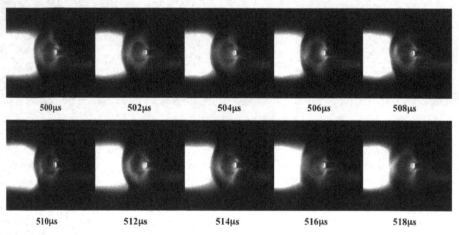

Fig. 13. High-speed photos of Magsail's magnetosphere during its quasi-steady operation; time corresponds to elapsed time after discharge of SWS is initiated. (J_{sws}=11.6 kA, J_{coil}=2 kA).

This magnetospheric fluctuation is characterized by high-speed photography based on the fact that the location of magnetopause corresponds to the dark regions in photos. Figure 14 shows the fluctuation of magnetospheric size for two plasma parameters (the discharge currents of SWS are 7.1 kA and 11.6 kA). The magnetospheric size is defined as a distance from coil center to the dark region. The averaged magnetospheric size is 145 mm in Fig.14a), and 80 mm in Fig. 14b). The magnetospheric size for J_{sws}=11.6 kA is larger than the case for J_{sws} =7.1 kA by about 10 mm, because the dynamic pressure of the simulated solar wind is smaller. The amplitudes of these fluctuations are about 10 mm, which corresponds to the fluctuation of thrust of about 25%. The power spectrum densities show that the dominant

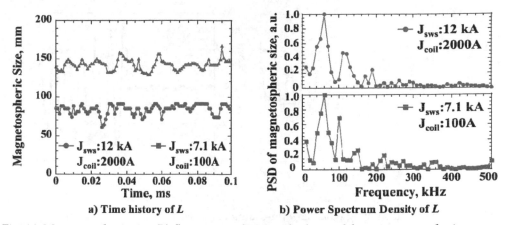

a) Time history of _L_   **b) Power Spectrum Density of _L_**

Fig. 14. Magnetospheric size (L) fluctuation of Magsail; obtained from image analysis.

frequency was about 60 kHz. This frequency corresponds to a natural mode of bouncing magnetosphere. Such a bouncing magnetosphere is also expected in space in a frequency range of 1-10 Hz in the case of moderately sized MagSail (L~100 km).

4.2 Thrust measurement of MagSail

The magnetic cavity is blocking the plasma flow emitted from the MPD arcjet to produce a force exerting on a miniature MagSail spacecraft (coil). To evaluate this solar wind momentum to thrust conversion process, impulse measurements were carried out by the parallelogram-pendulum method (Ueno, et al., 2009; Ueno, et al., 2011). For the measurement, a coil simulating MagSail was mounted on a thrust stand suspended with four steel wires as shown in Fig.6.

The impulse of a Magnetic Sail is given by the following equation:

$$\left(F\Delta t\right)_{MagSail} = \left(F\Delta t\right)_{Total} - \left(F\Delta t\right)_{SWS} \tag{11}$$

When only the solar wind simulator is operated, the pressure on the coil surface produces impulse; this impulse corresponds to $(F\Delta t)_{SWS}$ in Eq. (5). If the coil current is initiated during the solar wind operation, the impulse, $(F\Delta t)_{Total}$, becomes larger than $(F\Delta t)_{SWS}$. Thrust by a Magnetic Sail is defined as the difference between the two impulses divided by the SWS operation duration (Δt = 0.8 ms):

$$F_{MagSail} = \frac{\left(F\Delta t\right)_{MagSail}}{\Delta t} \tag{12}$$

In the experiment, the displacement of the pendulum was measured with a laser position sensor. For the calibration of the pendulum and position sensor combination, impulses of known magnitude were applied to the coil simulating MagSail.

Displacement waveforms of the thrust stand were observed as in Fig.15 when a coil was immersed into the plasma flow. We can see that the maximum displacement for 1.1-kA coil

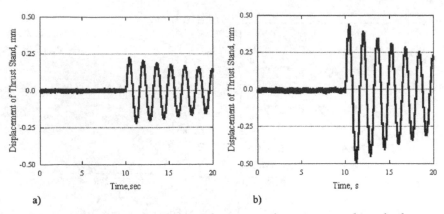

a)   b)

Fig. 15. Thrust stand's swing when 25-mm-diameter coil was immersed into hydrogen plasma flow (u_{sw}=47 km/s, n=1.8x10^{19} m^{-3}); a) without coil current, and b) 1.1 kA coil current (Simulator was initiated at 10s).

current in Fig.15b) is about two times larger than that without a coil current in Fig.15a). Figure 16 shows thrust data of MagSail for various magnetic moments where the magnetic moments were derived from the coil geometry and the coil current. It was confirmed that the thrust level is increased when increasing the magnetic moment of coils and thrust is proportional to $(M_m)^{2/3}$, which is consistent with eq.(5) because $L \propto (M_m)^{1/3}$.

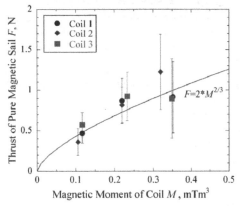

Fig. 16. Thrust vs. magnetic moment in the case of MagSail; SWS was operated for hydrogen plasma flow (u_{sw}=47 km/s, n=1.8x10^{19} m^{-3}), and three types of coils (radius=25 mm (coil 1,2) or 35 mm (coil 3)) are positioned at X=0.6 m.

If ions in a solar wind plasma penetrate into the magnetic cavity, the ions experience Larmor gyration. As was shown in Figs.4 and 5, if $r_{Li}<L$ at magnetospheric boundary, ions are reflected, but if $r_{Li}>L$, ions will not be reflected at the magnetospheric boundary but will penetrate deep into the magnetic cavity without producing thrust by MagSail. As a result, small $r_{Li}<L$ is required to obtain a significant thrust level. Such trend can be summarized in Fig.17, in which C_d (non-dimensional thrust) becomes smaller at a higher r_{Li}/L value.

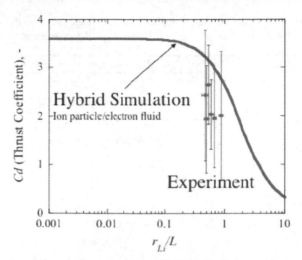

Fig. 17. Non-dimensional thrust characteristics of MagSail; C_d for various r_{Li}/L.

Through the experiment of MagSail, MHD to ion scale MagSail (r_{Li}/L=0.1~1.0) was demonstrated in laboratory.

Seeing the detail of Fig.17, however, one can see that laboratory experiment shows rather small thrust level than that by numerial simulation. This discrepancy is caused as a result of the collisional effect that is inevitable in laboratory [Kajimura et al., 2010], and it means that the laboratory experiment cannot predict the entire feature of collisionless plasmas in space. In the laboratory experiment, the momentum of a plasma flow in a vacuum chamber decreases as a result of collisions with neutral particles, leading to a decrease in thrust level.

4.3 Checking the similarity law

As described before, laboratory experiment is prepared so that the plasma flow follows the similarity law of MagSail/MPS in space. In the following, similarity law including the collisional effect in laboratory is checked based on measured plasma parameters.

When the hydrogen propellant is ionized, protons without a bounded electron are produced. Most of the plume plasma jet released from the SWS-MPD consists of atoms and protons, but molecules are minor species. By estimating the degree of ionization, the scaling parameters in our scale model are to be discussed.

From the data in Table.1, the radius of the plume plasma was 0.35 m at the coil position (X=0.7 m), which indicates that the plume divergence angle was 45 degree. The mass flow rate of a plasma jet from the MPD arcjet is the sum of a neutral flow rate and an ion flow rate,

$$\dot{m} = \left(m_n n_n u_n + m_i n_i u_i \right) A \tag{13}$$

From eq.(13), n_n=1.7x10^{19} m^{-3} is obtained for $\dot{m}$ =0.4 g/s, u_n=u_i=45 km/s, and the cross-sectional area of SWS's plume A= π (0.35)^2 m^2. In a realistic MPD arcjet flow, however, the velocity slip between the ions and the atoms are usually found. Including this effect, n_n=3.4x10^{19} m^{-3} for u_i=45 km/s, and u_n= 22.5 km/s. This estimation gives ionization ratio

α=0.2. Using the obtained species concentrations, the mean free paths can be calculated, and an example is shown in Table 3 for typical operational parameters of SWS.

Variable		Expression or assumption	Value
Electron density	n_e	measured typical value	2.0×10^{18} m^{-3}
Electron temperature	T_e	measured typical value	11640 K
Ion/atom temperature	T_i	$T_i \approx T_e$ is assumed	11640 K
Ion velocity	u_i	measured typical value	45 km/s
Thermal velocity	$u_{th,e}$	$\sqrt{8kT_e/m_e\pi}$	6.7×10^5 m/s
	$u_{th,i}\left(\approx u_{th,n}\right)$	$\sqrt{8kT_i/m_i\pi}$	1.6×10^4 m/s
Size of magnetic cavity	L	design value	0.1 m
Mean free path	λ_{ei}	$u_{th,e}/\nu_{ei}$	0.012 m
	λ_{en}	$1/n_n\sigma_{en}$	0.074 m
	λ_{nn}	$1/\sqrt{2}n_n\pi d^2$	0.60 m
Collision frequency	ν_{ei}	$3.64\times10^{-6}n_e\ln\Lambda/T_e^{3/2}$	5.4×10^7 /s
	ν_{en}	$u_{th,e}/\lambda_{en}$	9.1×10^6 /s
	ν_{nn}	$u_{th,n}/\lambda_{nn}$	2.6×10^4 /s
Electric conductivity	σ	$n_e e^2/m_e\left(0.51\nu_{ei}+\nu_{en}\right)$	2040
Non-dimensional variables		Expression	Value
Knudsen Number	(electron-ion)	λ_{ei}/L	0.12
	(electron-atom)	λ_{en}/L	0.74
	(atom-atom)	λ_{nn}/L	6
Magnetic Reynolds Number	Rm	$\sigma\mu_0 u_i L$	12

Table 3. Operational, measured, and scaling parameters of MagSail in laboratory.

In this experiment, coil diameter (18 mm) and magnetospheric size ($L\sim$100 mm) are smaller than the mean free path of hydrogen atoms, hence neutral flow behaves as collisionless particles. Proton and hydrogen atoms also behave as collisionless particles, but electron-heavy particle collision is significant and hence the mean free path of electrons is smaller than L. Let's consider electron motions in laboratory experiment. An electron coming into the magnetosphere is trapped in the magnetosphere, and then it makes a mirror movement in a magnetic flux tube. During such a mirror motion, an electron moves to another magnetic flux tube as a result of a collision with a heavy particle, or completely released from the magnetosphere. Such diffused electron motions will change the skin depth of a magnetopause from δ to δ_D (Bachynski & Osborne, 1965)

$$\delta_D = \frac{c^2\left(0.51\nu_{ei}+\nu_{en}\right)}{\omega_p u_{sw}} \tag{14}$$

It is derived that the ratio of δ_D to L is equal to the reciprocal of Rm as

$$\frac{\delta_D}{L} = \frac{1}{Rm} \tag{15}$$

Also, converting the mean free path data to collision frequencies, electric conductivity is evaluated as

$$\sigma = \frac{e^2 n_e}{m_e \left(0.51 v_{ei} + v_{en} \right)} \tag{16}$$

where v_{ei} is electron-ion collision frequency and v_{en} is electron-neutral collision frequency. Using the values in Table.3, σ is calculated as 2040/ Ω m; then Rm=12 is obtained for u_i=45 km/s and L=0.1m. In this case, the conditions for Rm (Rm>>1) and δ_D /L <<1 are satisfied.

By further tuning SWS (by increasing Rm) more suitable plasma flow for MagSail in laboratory might be provided. To further enlarge Rm, n_n should be decreased and T_e should be increased (see the equations in Table.3).

5. Extension of MagSail: M2P2 and magnetoplasma sail

5.1 The principle of M2P2 and magnetoplasma sail

For a MagSail in space, we must create a large magnetosphere to produce a significant thrust level because the dynamic pressure of the solar wind is very small even around the Earth. Using eqs.(4) and (5), thrust level of Magsail F is rewritten as

$$F = C_d \frac{1}{4} \left(\rho u_{sw}^2 \right)^{5/3} \left(\pi \mu_0 \right)^{1/3} \cdot M_m^{2/3} \tag{17}$$

This equation shows that F is proportional to $(M_m)^{2/3}$, where $M_m = \mu_0 n J \pi r_c^2$ is the magnetic moment. Hence, a coil with a large diameter (r_c), a high current (J), or many turns (n) are required to obtain a high thrust level. In the case of MagSail, a large magnetosphere might be formed using a coil of several tens of kilometer in diameter; for example, Zubrin proposed 10-N-class MagSail spacecraft equipped with a 64-km-diameter coil. This design is, however, impractical at the current technology level.

In order to overcome this issue, the idea to make a large magnetosphere by a compact coil diameter (~several meters) with a plasma jet was proposed instead of employing a large-scale coil. This propulsion system, illustrated in Fig.18, is called mini-magnetospheric plasma propulsion (M2P2) or MagnetoPlasma Sail (MPS). This idea gained great interest because the study by Winglee et al. (Winglee, et al., 2001) suggested the possibility of dramatically large acceleration by M2P2 spacecraft in deep space.

The M2P2/MPS sail uses an artificially generated magnetic field, which is inflated by the injection of high- or low-energy plasma. As shown in Fig.19, based on either magnetic field inflation concept either by high-velocity plasma jet or by equatorial ring-current, either of which initiates additional plasma current inside a magnetosphere to increase the magnetic moment of the system. Magnetic field inflation by plasma jet is carried out when a collisionless plasma jet is introduced in a magnetic field, and magnetic field is frozen into the plasma flow so that the plasma flow expands the magnetic field. In contrast to the method

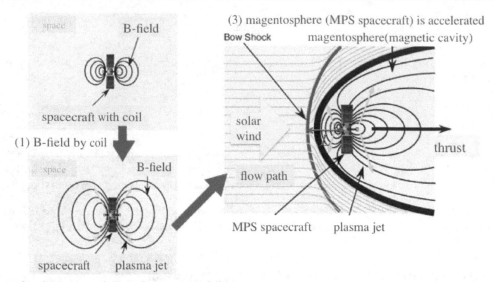

(1) B-field by coil

(2) plasma jet to inflate the magnetic field

Fig. 18. M2P2/MPS concept.

using a radial flow continuously flowing out, when a low-energy plasma is trapped in the magnetosphere, equatorial ring-current is initiated due to drift motions of both ions and electrons. These plasma currents may allow the deployment of the magnetic field in space over large distances (comparable to those of the MagSail) with the B-field strengths that can be achieved with existing technology (i.e., conventional electromagnets or superconducting magnets). Additionally, one potential significant benefit of the M2P2/MPS would be its small size of the hardware (even though the magnetic bubble is very large); this would eliminate the need for the deployment of large mechanical structures that are presently envisaged for MagSail or solar light sails (Frisbee, 2003, Yamakawa, et al., 2006).

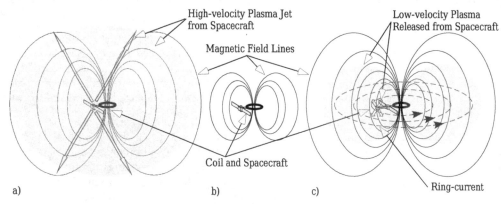

Fig. 19. Magnetic field inflation concept by high-velocity plasma jet or by equatorial ring-current.

However, feasibility of the M2P2/MPS sail concept is not established yet (Khazanov, et al., 2005; Omidi & Karimabadi, 2003}. From the physical point of view, we have questions on the principle of thrust production: under what operational condition is the M2P2/MPS concept valid?; is efficient thrust production really possible for M2P2/MPS? These are still open questions. Also, from the spacecraft design point of view, it is not known that if M2P2/MPS obtains much higher acceleration than that of existing technology such as electric propulsion or solar sail. We are continuing our efforts on theoretical and experiment studies to reveal these issues.

5.2 Preliminary experiment of M2P2/MPS

In contrast that a lot of experiments revealed the physics and thrust characteristics of MagSail, only a few data are available in the case of M2P2/MPS. The first experiment on the magnetic field inflation concept was tried by Winglee and his group (Winglee, et al., 2001; Ziemba, et al., 2001; Slough, et al., 2001; Ziemba, et al., 2003; Giersch, et al., 2003). A high-density helicon or an arc plasma source is located in a drum-like solenoid to observe magnetic field inflation. The demonstration was conducted in a space chamber of 1 m in diameter. He also tried full demonstration of M2P2. However, due to facility limitaion, they cannot demonstrate some important features of M2P2; for example, 1) the entire interaction of M2P2 is not tested due to limited diameter of the solar wind simulator, and 2) what amount of momentum transfers from the solar wind to the spacecraft (coil) are not validated.

Scaled down model approach by our group with a very small plasma source from the center of the coil seems reasonable to simulate the whole system of a M2P2/MPS sail. In Fig.20, an experimental system for MPS is depicted, which consists of a high-power magnetoplasmadynamic (MPD) solar wind simulator (MPD_SWS) and a miniature MPS spacecraft (Funaki, et al., 2007, Ueno, et al., 2009). Close-up views around the miniature MPS spacecraft (scale model) are shown in Fig.21. The miniature MPS spacecraft has a solenoidal coil and MPD arcjets (MPD_Inf, 20-mm-diameter small MPD arcjets) for plasma jet injection. All of these devices (the MPD_SWS, the MPD_Inf, and the coil) are operated again in a quasi-steady mode of about 1 ms duration. To describe the physics of the magnetic field inflation process, the most important parameter is r_{Li}/L_{inf}. In this parameter, L_{inf} is the frozen-in point where local β value is unity, so the B-field inflation occurs. Strictly speaking, L_{inf} is defined as a distance from the coil center to a typical frozen-in point. Also, the B-field inflation process requires small ion gyration radius when the B-field inflation is possible in the MHD approximation. Hence, two inequalities, $r_{Li}/L_{inf} < 1$ and $L_{inf} < L$, are added to the inequalities in Table.1. In our preliminary experiment, the miniature MPS spacecraft with a coil of 50-75 mm in diameter was located in a downstream position of the MPD arcjet (SWS) to produce up to 2-T magnetic field at the center of the coil. Into this magnetic field produced by the miniature MPS spacecraft, a plasma jet from the MPD_SWS was introduced to see possible interaction.

Typical snapshot of MPS experiment is shown in Figs.22 and 23. It is obvious that the incoming solar wind plasma flow is blocked by the magnetic cavity in both the MagSail mode as shown in Fig.23(a) and in the miniature MPS mode in Fig.23(b). Comparing these photos, it is observed that the magnetosphere is inflated from an arched shape in Fig.23(a) to a rather flat shape in Fig.23(b) by the plasma injections from inside the solenoidal coil. The

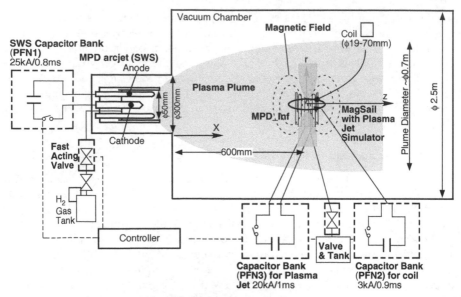

Fig. 20. Experimental setup for magnetoplasma sail (MPS).

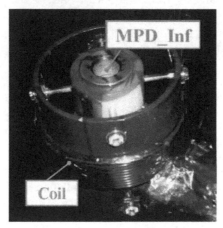

Fig. 21. Miniature MPS spacecraft (20-mm diameter MPD is installed inside 75-mm-diameter coil).

distances from the coil center to the magnetospheric boundary are 119.3 mm in Fig.23(a), and 145.3 mm in Fig.23(b); these distances are estimated from each photo. Therefore, from this evaluation, the magnetic field expansion observed is 26 mm. This is our first and successful magnetic field inflation experiment in laboratory.

Currently, experimental efforts are going on: 1) to form a large MPS magnetosphere, and 2) to detect increased thrust level by MPS operation. Direct thrust measurements of the miniature MPS is very interesting, but due to facility limitation, thrust measurement so far was not successful. Our goal is to evaluate a clear thrust gain (thrust of MPS divided by

Fig. 22. Typical flow around miniature MPS during quasi-steady operation (discharge current of SWS is 8 kA (H_2;0.4 g/s), and discharge current of MPD_Inf is 4.2 kA (H_2;0.05 g/s)).

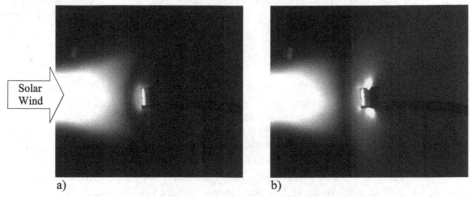

a)   b)

Fig. 23. Photo of scale-model experiment of MPS; a) miniature MagSail and b) miniature MPS. (simulated solar wind is introduced to 5cm/20turn coil. SWS: u_{sw}=45 km/s, n_{sw}=1.8x10^{19} m^{-3}, H_2, 0.4 g/s. From inside the coil, a plasma jet is ejected from 2-cm diameter MPD : H_2, 0.04 g/s, 0.06 T at the surface of the MPD, and J=3.5 kA for the miniature MPS).

thrust of MagSail) that is possible by MPS. For this purpose, a higher power plasma source is going replace the small (20-mm-diameter) MPD arcjet to obtain two times larger magnetic cavity size than that of MagSail. Also, several plasma sources are prepared that offer both low velocity and high velocity plasma jet. These setups will enable thrust gain more than 200%, so that an increased thrust level might be detectable.

6. Conclusion

Starting from the basic theory of sail propulsion using the solar wind, current status of solar wind sail (MPS) research, in particular experiment using a plasma wind tunnel, is described. In the experiment, using a self-field magnetoplasmadynamic arcjet as a solar wind simulator, quasi-steady interaction between a plasma flow and a magnetic cavity is obtained for a Magsail (consisting of only a coil) and for an MPS (which accomplishes an inflated magnetic cavity by a plasma jet from inside a coil). So far, thrust measurement of these sails was conducted for only a few limited cases of Magsail (without plasma jet from spacecraft), and extension of the experimental set-up to MPS thrust measurement is still going on. During the quasi-steady interaction of about 0.8 ms, the dynamic behavior Magsail/MPS in laboratory is observed, and it was found that the magnetic cavity of Magsail is fluctuating in a frequency rage of 10-800 kHz. Thrust by Magsail is hence produced under rather turbulent plasma flow to magnetic cavity interaction.

Inspired by the ideas of MagSail and its derivatives, many new sail propulsion system are advocated. Khazanov and Akita proposed the usage of solar radiation to enhance thrust by MagSail (Khazanov et al., 2005 and Akita and Suzuki, 2004). Also, Slough proposed a way to capture the solar wind momentum by using a rotating magnetic field, and the new system is called 'the plasma magnet' (Slough, 2005). The plasma magnet is already accessed by laboratory experiment (Slough, 2007). In 2004, Janhunen proposed a completely new idea using the solar wind, which has a set of thin long wires being kept at a high positive potential so that the wires repel and deflect incident solar wind protons and it is called as electrostatic sail (Janhunen, 2004; Janhunen & Sandroos, 2007). These ideas are not so matured to proceed to a flight demonstration in space, but attractive solar wind sails that are competitive against the existing thruster technology is emerging.

7. Acknowledgment

We would like to express our acknowledgments to the members of the MPS working group (in particular, Dr. Kazuma Ueno, Mr. Yuya Oshio for providing experimental data, and Dr. Hiroyuki Nishida, Dr. Yoshihiro Kajimura, Dr. Iku Shinohara, Dr. Masaharu Matsumoto, Mr. Yasumasa Ashida, Dr. Hirotaka Otsu, Dr. Kazuhisa Fujita, and Prof. Hideyuki Usui) for their valuable activities in the analysis of MPS. This research is supported by: 1) the Grant-in-Aid for Scientific Research (A) (No.21246126) of the Japan Society for Promotion of Science, 2) the space plasma laboratory and the engineering committee of the institute of space and astronautical science in Japan Aerospace Exploration Agency (JAXA), and 3) Japan Science and Technology Agency, CREST.

8. References

Akita, D., and Suzuki, K., On the Possibility of Utilization of Radiation Pressure in Magnetic Sails, Proceedings of Symposium on Flight Mechanics and Astrodynamics, (2004), pp.1-4 (in Japanese)

Andrews, D.G., and Zubrin, R.M., Magnetic Sails and Interstellar Travel, Journal of the British Interplanetary Society, Vol.43, (1990), pp.265-272, ISSN 0007-084X

Bachynski, M.P. and Osborne, F.J.F., Laboratory Geophysics and Astrophysics, in Advances in Plasma Dynamics (Anderson, T.P. and Springer R.W. Eds), (1965). Northwestern University Press

Fujita, K., Particle Simulation of Moderately-Sized Magnetic Sails, The Journal of Space Technology and Science, Vol.20, No.2, (2004), pp.26-31, ISSN 0911-551X

Funaki, I. Kojima, H., Yamakawa, H., Nakayama, Y., and Shimizu, Y., Laboratory Experiment of Plasma Flow around Magnetic Sail, Astrophysics and Space Science, Vol.307, No.1-3, (2007), pp.63-68, ISSN 0004-640X

Funaki, I., Kimura, T., Ueno, K., Horisawa, H., Yamakawa, H., Kajimura, Y, Nakashima, H., and Shimizu, Y., Laboratory Experiment of Magnetoplasma Sail, Part 2: Magnetic

Field Inflation, Proceedings of 30th International Electric Propulsion Conference, IEPC2007-94, Florence, Sept. 2007

Frisbee, R. H., Advanced Space Propulsion for the 21st Century, Journal of Propulsion and Power, Vol.19, No.6, (2003), pp.1129-1154, ISSN 0748-4658

Giersch, L., Winglee, R.M., Slough, J., Ziemba, T., Euripides, P., Magnetic Dipole Inflation with Cascaded Arc and Applications to Mini-magnetospheric Plasma Propulsion, 39th AIAA/ASME/SAE/ASEE Joint Propulsion Conference & Exhibit, AIAA-2003-5223, Huntsville, July 2003

Janhunen, P., Electric Sails for Spacecraft Propulsion, Journal of Propulsion and Power, Vol.20, No.4, (2004), pp.763-764, ISSN 0748-4658

Janhunen, P., and Sandroos, A., Simulation Study of Solar Wind Push on a Charged Wire: Basis of Solar Wind Electric Sail Propulsion, Annales Geophysicae, Vol.25, (2007), pp.755–767, ISSN 0992-7689

Kajimura, Y., Usui, H., Ueno, K., Funaki, I., Nunami, M., Shinohara, I., Nakamura, M., and Yamakawa, H., Hybrid Particle-in-Cell Simulations of Magnetic Sail in Laboratory Experiment, Journal of Propulsion and Power, Vol.26, No.1, (2010), pp.159-166, ISSN 0748-4658

Khazanov, G., Delamere, P., Kabin, K., Linde, T. J., Fundamentals of the Plasma Sail Concept: Magnetohydrodynamic and Kinetic Studies, Journal of Propulsion and Power, Vol.21, No.5, (2005), pp.853-861, ISSN 0748-4658

Nishida, A. Ed., Magnetospheric Plasma Physics, (1982). Center for Academic Publications Japan, ISBN90-277-1345-6

Obayashi, T., Solar Terrestrial Physics (1970). Syokabo, Tokyo (in Japanese), ISBN4-7853-2405-8

Omidi, N., and Karimabadi, H., Kinetic Simulation/Modeling of Plasma Sails, 39th AIAA/ASME/SAE/ASEE Joint Propulsion Conference and Exhibit, AIAA-2003-5226, Huntsville, July 2003

Oshio, Y., Ueno, K., and Funaki, I., Fluctuation of Magnetosphere in Scale-model Experiment of Magnetic Sail, 31st International Electric Propulsion Conference, IEPC2007-94, Michigan, Sept. 2009

Slough, J., High Beta Plasma for Inflation of a Dipolar Magnetic Field as a Magnetic Sail (M2P2), Proceedings of 27th International Electric Propulsion Conference, IEPC01-202, Pasadena, Oct. 2001

Slough, J. and Giersch, L., The Plasma Magnet, 41st AIAA/ASME/SAE/ASEE Joint Propulsion Conference & Exhibit, AIAA-2005-4461, Tucson, Arizona, July 2005

Slough, J., Plasma Sail Propulsion based on the Plasma Magnet, Proceedings of 30th International Electric Propulsion Conference, IEPC2007-15, Florence, Sept. 2007

Ueno, K., Kimura, T., Ayabe, T., Funaki, I., Yamakawa, H., and Horisawa, H., Thrust Measurement of Pure Magnetic Sail, Transactions of JSASS, Space Technology Japan, Vol. 7, (2009), pp.Pb_65-Pb_69, ISSN 1884-0485

Ueno, K., Funaki, I., Kimura, T., Horisawa, H., and Yamakawa, H., Thrust Measurement of Pure Magnetic Sail using Parallelogram-pendulum Method, Journal of Propulsion and Power, Vol.25, No.2, (2009), pp.536-539, ISSN 0748-4658

Ueno, K., Oshio, Y., Funaki, I., Ayabe, T., Horisawa, H., and Yamakawa, H., Characterization of Magnetoplasma Sail in Laboratory, Proceedings of 27th

International Symposium on Space Technology and Science, ISTS-2009-b-42, Tsukuba, June 2009

Winglee, R.M., Slough, J., Ziemba, T., and Goodson, A., Mini-Magnetospheric Plasma Propulsion: Tapping the Energy of the Solar Wind for Spacecraft Propulsion, Journal of Geophysical Research, Vol.105, No.21, (2000), pp.21067-21078, ISSN0148-0227

Winglee, R.M., Ziemba, T., Euripides, P., and Slough, J., Computer Modeling of the Laboratory Testing of Mini-Magnetospheric Plasma Propulsion, Proceedings of 27th International Electric Propulsion Conference, IEPC-01-200, Pasadena, Oct. 2001

Winglee, R.M., Euripides, P., Ziemba, T., Slough, J., and Giersch, L., Simulation of Mini-Magnetospheric Plasma Propulsion (M2P2) Interacting with an External Plasma Wind, 39th AIAA/ASME/SAE/ASEE Joint Propulsion Conference & Exhibit, AIAA-2003-5224, Huntsville, July 2003

Yamakawa, H., Funaki, I., Nakayama, Y., Fujita, K., Ogawa, H., Nonaka, S., Kuninaka, H., Sawai, S., Nishida, H., Asahi, R., Otsu, H., and Nakashima, H., Magneto Plasma Sail: An Engineering Satellite Concept and its Application for Outer Planet Missions, Acta Astronautica, Vol.59, (2006), pp.777-784, ISSN 0094-5765

Ziemba, T.M., Winglee, R.M., Euripides, P., and Slough J., Parameterization of the Laboratory Performance of the Mini-Magnetospheric Plasma Propulsion (M2P2) Prototype, Proceedings of 27th International Electric Propulsion Conference, IEPC-01-201, Pasadena, Oct. 2001

Ziemba, T., Euripides, P., Winglee, R.M., Slough, J., Giersch, L., Efficient Plasma Production in Low Background Neutral Pressures with the M2P2 Prototype, 39th AIAA/ASME/SAE/ASEE Joint Propulsion Conference & Exhibit, AIAA-2003-5222, Huntsville, July 2003

Zubrin, R.M., and Andrews, D.G., Magnetic Sails and Interplanetary Travel, Journal of Spacecraft and Rockets, Vol.28, No.2, (1991), pp.197-203, ISSN 0022-4650

Permissions

The contributors of this book come from diverse backgrounds, making this book a truly international effort. This book will bring forth new frontiers with its revolutionizing research information and detailed analysis of the nascent developments around the world.

We would like to thank Marian Lazar, for lending his expertise to make the book truly unique. He has played a crucial role in the development of this book. Without his invaluable contribution this book wouldn't have been possible. He has made vital efforts to compile up to date information on the varied aspects of this subject to make this book a valuable addition to the collection of many professionals and students.

This book was conceptualized with the vision of imparting up-to-date information and advanced data in this field. To ensure the same, a matchless editorial board was set up. Every individual on the board went through rigorous rounds of assessment to prove their worth. After which they invested a large part of their time researching and compiling the most relevant data for our readers. Conferences and sessions were held from time to time between the editorial board and the contributing authors to present the data in the most comprehensible form. The editorial team has worked tirelessly to provide valuable and valid information to help people across the globe.

Every chapter published in this book has been scrutinized by our experts. Their significance has been extensively debated. The topics covered herein carry significant findings which will fuel the growth of the discipline. They may even be implemented as practical applications or may be referred to as a beginning point for another development. Chapters in this book were first published by InTech; hereby published with permission under the Creative Commons Attribution License or equivalent.

The editorial board has been involved in producing this book since its inception. They have spent rigorous hours researching and exploring the diverse topics which have resulted in the successful publishing of this book. They have passed on their knowledge of decades through this book. To expedite this challenging task, the publisher supported the team at every step. A small team of assistant editors was also appointed to further simplify the editing procedure and attain best results for the readers.

Our editorial team has been hand-picked from every corner of the world. Their multi-ethnicity adds dynamic inputs to the discussions which result in innovative outcomes. These outcomes are then further discussed with the researchers and contributors who give their valuable feedback and opinion regarding the same. The feedback is then collaborated with the researches and they are edited in a comprehensive manner to aid the understanding of the subject.

Apart from the editorial board, the designing team has also invested a significant amount of their time in understanding the subject and creating the most relevant covers. They scrutinized every image to scout for the most suitable representation of the subject and create an appropriate cover for the book.

The publishing team has been involved in this book since its early stages. They were actively engaged in every process, be it collecting the data, connecting with the contributors or procuring relevant information. The team has been an ardent support to the editorial, designing and production team. Their endless efforts to recruit the best for this project, has resulted in the accomplishment of this book. They are a veteran in the field of academics and their pool of knowledge is as vast as their experience in printing. Their expertise and guidance has proved useful at every step. Their uncompromising quality standards have made this book an exceptional effort. Their encouragement from time to time has been an inspiration for everyone.

The publisher and the editorial board hope that this book will prove to be a valuable piece of knowledge for researchers, students, practitioners and scholars across the globe.

List of Contributors

A.G. Tlatov
Kislovodsk Mountain Station of the Central Astronomical Observatory of RAS at Pulkovo, Russia

B.P. Filippov
Pushkov Institute of Terrestrial Magnetism, Ionosphere and Radio Wave Propagation, Russian Academy of Sciences, Troitsk, Moscow Region, Russia

Karel Kudela
Institute of Experimental Physics, Slovak Academy of Sciences, Kosice, Slovakia

Natalia Buzulukova
NASA Goddard Space Flight Center/CRESST/University of Maryland College Park, USA

Mei-Ching Fok and Alex Glocer
NASA Goddard Space Flight Center, USA

Peter Stauning
Danish Meteorological Institute, Denmark

U. Villante and M. Piersanti
Dipartimento di Fisica, Università e Area di Ricerca in Astrogeofisica, L'Aquila, Italy

Maria S. Pulinets and Svetlana S. Znatkova
Skobeltsyn Institute of Nuclear Physics, Moscow State University, Moscow, Russia

Elizaveta E. Antonova, Maria O. Riazantseva and Igor P. Kirpichev
Skobeltsyn Institute of Nuclear Physics, Moscow State University, Moscow, Russia
Space Research Institute RAS, Moscow, Russia

Marina V. Stepanova
Physics Department, Universidad de Santiago de Chile, Chile

Oleg Troshichev
Arctic and Antarctic Research Institute, Russia

Ikkoh Funaki
Japan Aerospace Exploration Agency, Japan

Hiroshi Yamakawa
Kyoto University, Japan

9 781632 395726